I0055725

Advanced Guide to Bladesmithing

Forge Pattern Welded Damascus Swords, Japanese Blades, and Make Sword Scabbards

© **Copyright 2019 - All rights reserved.**

The content contained within this book may not be reproduced, duplicated or transmitted without direct written permission from the author or the publisher.

Under no circumstances will any blame or legal responsibility be held against the publisher, or author, for any damages, reparation, or monetary loss due to the information contained within this book, either directly or indirectly.

Legal Notice:

This book is copyright protected. This book is only for personal use. You cannot amend, distribute, sell, use, quote or paraphrase any part, or the content within this book, without the consent of the author or publisher.

Disclaimer Notice:

Please note the information contained within

this document is for educational and entertainment purposes only. All effort has been executed to present accurate, up to date, and reliable, complete information. No warranties of any kind are declared or implied. Readers acknowledge that the author is not engaging in the rendering of legal, financial, medical or professional advice. The content within this book has been derived from various sources. Please consult a licensed professional before attempting any techniques outlined in this book.

By reading this document, the reader agrees that under no circumstances is the author responsible for any losses, direct or indirect, which are incurred as a result of the use of information contained within this document, including, but not limited to, errors, omissions, or inaccuracies.

Table of Contents

Introduction

Do you want to make a knife? It might sound strange to some people, and they might even wonder why they would want to go through such stress when one can easily get inexpensive yet functional knives in stores. I've had the opportunity to be asked a question like this. Some people need a knife that could be in their pocket that they could just use to clean their fingernails and maybe open envelopes; others might be about them getting a hunting knife which they can carry around to spilled-out game if possible so it might not be able owning an expensive knife. Well, a lot more people see a knife as a working tool which should be cared for and, of course, a tool one should have. These people need a knife that is long-lasting; they need a knife that has an edge and can be sharpened easily.

People in the last category can hardly find knives in the stores worthy for them, and the make of the knives they see might not be up to their taste. So, they feel their best bet is making their knives. Making their own is just the best way they can have the kind of quality knife they desired.

For those that want to learn how to make knives and swords, this book is for you. You will be shown how to make them and the stages that are involved and will help save you time from getting inferior knives. Even the advanced stage of the blade markers art such as special parts, fittings and patterns will be discussed.

The good thing is that the materials needed are easily accessible and they are inexpensive. Each knife that you make can be a beauty and a work of art. You will be wowed at the kind of knife you will be able to produce. With this book, they will look smooth and beautiful, having built each with your personal touch. Another good thing about making your blade is that you make the selection of the steel personally so that it will meet what you like so that after heat-treating the blade you can draw the blade, i.e., the temper to meet your specifications. This particular book will give you all the information that you will need to follow.

Chapter 1. Pattern Welding

Pattern welding can be seen as a unique way of joining a sword blade from iron and steel parts. Many types of iron and steel parts are welded in the fire in such a way that a satisfying pattern can be seen on one or both sides of the knife blade. This pattern comes to be because the two used kinds of iron and steel laid in between the surface shone the light differently, in a particular way after some special polishing and processing. You must use some unique shapes of the parts to be welded to have and create a specific pattern, and again different kinds of iron/steel must be used if one wants to see a visible pattern. This is to say that to succeed with this, structural and compositional piling must be done and should be done in such a way that it produces the kind of pattern one wants. So let look at the different words that relate to ways of making a sword.

Pattern welding: Fire welding makes use of different kinds of iron and steel in a way that a particular kind of pattern occurs on the finished blade. Putting together through fire welding some pieces of iron and steel in a more or less any kind of way will make a pattern on the finished blade. This pattern is often random and isn't intentional. The finished sword is never a pattern welded sword but one done through piling. Patterns will appear if different steels have been used and it will appear at the surface of the sword. It doesn't matter if the steels are automatically different; all that matters is that there are optically different. It is imperative that all the steels on both sides should show at the surface. This is to say that one kind is enclosed by the other kind. Like in producing Japanese blades, you can't possibly see it, and there might be no pattern on it so that the term piling can be used for this.

1. Pattern welded sword is in the class of composite swords.

2. Pattern welded swords are related to the term Damascus, and this often brings a lot of confusion.

Piling

Piling can easily be described as a big piece of iron or steel made by fire welding the smaller pieces. Piling can be done in a casual or complex way. Composite swords and pattern welded swords are readily produced through highly complex piling. Though, a swordsmith can do complex piling in both smart and foolish ways. The foolish way can be welding soft iron for the cutting edges to be hardcore. So, any pattern-welded sword is a sword that was produced by compositional and structural piling; but not all swords produced by compositional and structural piling are said to be a pattern-welded sword just like Japanese swords.

Damascus Technology

This term in history has been used for diverse things, and this has nothing to do with the location of Damascus. The term is said to be introduced because of some mistakes or misunderstanding.

Damascus Swords

Truthfully, no sword emanated from Damascus. Some people do not know this and might believe that a Damascus sword is a better one which has some pattern, though the property of having a very good sword is related to having a good pattern. It should be noted that pattern-welded swords are essential objects of art. Pattern wielding in its prime was never made because the method was reserved for good swords. One problem with this statement is that nobody has ever seen an old pattern-welded sword so magnificently made.

Traditional pattern-welded swords can be said to be 1,000 years old now. The old pattern-welded sword was trended in northern Europe at around 200 AD – 800 AD; but after the last years the slowly made their way out but really these complexities with pattern welding didn't just happen today. There has been a consistent development from piling to complex pattern welding. Earlier, some attempts at structural and compositional did create good swords, and on the side, too, there were simple patterns. Moving beyond this might only beautify the sword, not make the sword better.

How to Make Pattern Welding

Pattern welding uses butt welding. Four rods will be used, and more than ten might be needed. Each piece is 150 mm x 25 mm x 10 mm. Two rods are almost made of same hard steel and readily used for the blade's edges. Then, the rods that will create the pattern are produced from two different plates of steel. A standard that one faces is a package of seven layers, four made from soft steel, while the remaining three are made from hard steel. The hardness doesn't count though. What counts is the bright-dark contrast in the finished blade, not how hard it is. To get this effect, one can use phosphorus-rich and phosphorus-free iron, and this is the common approach in making what is called a striped rod iron, which contains enough phosphorus having a bright whitish appearance, while phosphorous-free iron or steel is dark. So, we will be using neutral terms bright and dark iron or steel. Coupling phosphorus-free carbon-lean and carbon-rich steel isn't a good idea because, during the high-temperature smithing needed for fire welding, the carbon concentration might or might not equalize by diffusion. At this point, there can't be a well-defined color change at the edges any longer. Also, since phosphorus spreads slower than carbon, the difference in phosphorus concentration isn't washed away.

We will be starting with producing a very simple pattern-welded blade with about three patterned striped rods. Some tips you need to know first about simple pattern-welded swords:

1. Get good enough raw material with these specifications

    a. Bright steel—make sure to get phosphorus-rich wrought iron because it is the best.

    b. Dark steel—a phosphorous-free

medium will do or whatever will have a variation to the bright steel.

c. Hard steel—needed for the edge, the steel should have at least eutectoid components about 0.7% carbon; the highest carbon concentration should be considered since one need more carbon during forging.

d. Other Materials/ Tools needed:

    i. Large File

    ii. Small file set

    iii. Lawnmower blade

    iv. Vise

    v. Angle Grinder discs (cutting, grinding, and finishing)

    vi. Angle Grinder

    vii. Borax

    viii. Canola Oil

    ix. Tongs

x.   Ferric Chloride

xi.   Anvil

xii.   Forge

xiii.   Drill (or drill press)

xiv.   Exotic   Wood   (Figured Walnut)

xv.   Drill Bits (1/8" and 5/32")

xvi.   Sandpaper   (220-1000 grit)

xvii.   Mini Sledge Hammers

xviii.   Saw (hacksaw)

xix.   1/8" Brass Pin

xx.   5/32" Rivet

xxi.   5-minute Epoxy

xxii.   Palm Sander

xxiii.   Rejuvenating Oil (or wood finish of choice)

xxiv.   Kydex Plastic (or flexible 1/8" plastic)

xxv.   Torch (optional)

xxvi. Scrollsaw (optional)

xxvii. Dremel (optional)

2. Faggot the three materials before you form bars; this means you should drive into a sheet, fold, and weld hammer into a sheet. You can repeat it up to ten times to ensure that the welds are in good condition and avoid putting dirt, influx inclusions, or oxide into the welding. Lastly, make two rods from the hard steel which should be as long as the blade to be, then make 9 – 12 plates from the dark and bright steel

3. Create three packages from three dark and four bright steel plates, i.e., the package, striped rod, and twist.

4. Try and fire weld each package and then drag it out to a rod about the same length as the blade-to-be. The rod's cross-section should be about 1 cm by 1 cm, and at the end, one will have what is called stripe rod. It seems to be the starting material for some pattern-welded swords, though not all.

5. Squeeze the striped rods evenly if you need a more complex pattern. This is to

ensure that the alignment of the layers in the non-squeezed regions is the same. You can also change the squeezed direction either clockwise or counterclockwise to have a squeezed striped rod.

6. The next thing to do is to place all rods next to one another. In the non-twisted regions, the layers should be perpendicular to the plane of the picture, so that a stripe pattern can be seen. Try to grind down the rods somehow most especially the flattening sides so it will be easy joining them. You can also try to fit in the squeezed part so that there are fitted like two screws having the same thread. This will mean that the pitch of the squeeze is the same with the two rods, though it might sound impossible. The remaining two hard steel rods will go outside if possible you can bend these rods a bit so that a snug fit is obtained at the end where the point will be.

7. Weld the assembly with fire, ensuring that the weld seams are perfectly done. Forge it into what is the basic sword shape if need be to provide a tang by

fire welding it, or better still, stretch the rods into the tang.

8. Create fuller by forging, if necessary. Grind the blade and let it have the cross-section feel. Remember that the pattern you will obtain will be dependent on how deep you grind into the squeezed rods.

9. This is the right time to quench-harden the edges; but before doing this, cover the body with protective mud for the differential hardening stage, making the cooling rate to be higher at the edge part than on the blade's body. By the way, this is a popular way of making Japanese swords with some specific hard edges even the ancestors are believed to have used the same method. At this point temper a bit like annealing at low temperatures after quenching. You can keep the time for the quenching short so that self-tempering can occur since the inside is still very much hot and heat up outside again after you might pull out the blade from the cold quenching liquid. This process can be called slag quenching.

10. Complete it by polishing the blade and forgetting the pattern exposed like etching it and allowing ample time for it to set. Add a good hilt to it, which will help add to the attraction and importance to the blade. By this you have completed a simple welded blade, you might be wondering at this point what a torsion Damascus pattern will look like? I'm longing to meet someone who can visualize what the distribution of dark and bright iron looks like in various depths of a twisted and striped rod. Maybe you can try figuring this out by yourself.

Remember that the front and back will look almost the same; what you see depends on how intense your grinding is and maybe on the chosen cross-sectional shape as schematically indicated. However, how deep you can grind depends a lot on the thickness of the blade. If a thickness of 5 mm is what you want in the center, on each side of a 10 mm rod, you may have to grind off 2.5 mm.

For swords, the ideal size for welding is about 125 x 40mm. Full sheets of 15N20 (1000 x 350mm) and 20C in sheets of 1450 x 260mm will work out to 24 sword pieces of 3mm x 130 x 40mm.

This is to say that you cannot get more flowery or circular patterns farther down the rod. Finally, you have made for yourself a simple yet impressive pattern sword; you might want to try out a more complicated one like having full freedom for patterns, maybe with different patterns on both sides of the blade and a fine pattern, too.

Basic Patterns

The layer count is just the starting point, but you can use less or more of it depending on your preference. In specialized Damascus patterns, such as radials or jellyrolls, fewer layers are needed. It is possible to form weld sections of the high and low-layer bars in a billet via patterning and get a high contrast. Patterning of the flat laminated billet can be achieved in diverse ways. The random pattern needs to be explained though. The layers remain very flat, and some disjointed sometimes occur during forging. Disjointed brings about flat layers to bend, and the result is a smooth organic look to the material. This is outstanding when the edge levels of the blade are finish-ground.

Twist patterns are a bar of the required

number of layers that is forged into a square having the corners forged down slightly. The bar gets heated until it's nearing the welding temperature and afterward twisted. This twisting can be slowly done or left tight for diverse effects. The center of each of these twists gives a star effect. Twisted blades are mandated to be left a little thicker than other patterns, as grinding deeper into it makes the star effect more efficient and the whole look more pleasing.

Ladder patterns can be achieved when you grind or press grooves across a Damascus bar. When the pattern is pressed into the blade, it undoubtedly doubles the thickness that was needed in the finished bar. The grooves are then pressed in with dies that are made of round rods and stop blocks that can be used to assist the normal thickness of the finished bar.

When the grooves have been pressed into the bar, it becomes ground flat thereby getting rid of all the high spots. Then, the bar is forged to the needed blade shape, and the pattern of the ladder becomes clear. If the ladder styles are crushed or milled into the bar, they become one-third of the bar thickness. Immediately, the grooves are grounded, the blade then is forged to size forging the whole grooves out of the bar bringing about a different ladder pattern. Now, whether the ladders are pressed or grounded, it should slant from side to side.

Making the raindrop or pool eye pattern is almost the same process as forge welding a ladder pattern, the only difference is the dimples pressed or drilled into the Damascus bar instead of the grooves. The result will be a pattern that resembles a bullseye or raindrops in a pool.

These are the main Damascus patterns and the same patterning techniques, and various others that are used for more complexed patterns. Before moving to more complex patterns and methods, the Damascus steelmaker should become skillful at the forge welding stage.

Complex Pattern Welded Blade

To get this pattern, you need to create eight striped rods, which should have 16 layers each. Then, grind down as deep as you can afford to get your needed pattern on a side and again, you grind down the backside so as to fire weld the thin rods to an extra core; that is four on the front side and another four on the backside. You will need about 64 each of the faggotted dark and bright iron plates for a start, in addition to the hard steel for the edges. What appears will be a sword that is created by piling with a veneer that produces the pattern. Well, the veneer might not all that be important as a sword properties it could just be there as additional material, it helps to hide all the mistakes from large slag inclusions, imperfect welding, and others flaws that might be visible on the made blade.

Making a sword of this magnitude is hard work and need skillful hands because just a mistake like a wrong bang on the hammer or a part of a millimeter not on during grinding can damage the blade. Most ancient sword bearers were mostly about fashion because one enjoys the glamour that comes with sporting a sword even those that can't afford do same. Though some pattern-welded swords had only one side patterned, the last gold hilt spatha that was discovered in 1970 had gold on the show side of the sword only like others. It is on record that this show off pattern-welded sword belonged to an Alemanni that live about 550 AD and near his home town. He was famous for using the plain sax of his sword to harm an opponent. If anyone thinks that the fashion attitude of time past is different from that of today, then he/she needs a rethink, everything has its own time same with a veneered pattern-welded sword. It was loose ground to a more radical new fashion like the ulfberht sword.

How to Create the "W's" Design

"W's" design and mosaic Damascus are other examples of improved designs. While creating the "W's" pattern, the initial billet is piled up like a flat-layered billet; then it is welded. In the lengthening process, the bullet is turned around to form right angles, after which it is hammered to form a rectangular bar with vertical layers. Scales are ground off the bar, cut into sections, and piled up again.

After the second weld arrangement is finished, the layers are still vertical. The bar is recut and piled up again. Scraping the ends of these pieces uncovers the contorted layers are already taking the 'W' form. The third arrangement will further contort the layers and create striking "W's." This pattern is suitable for any number of layers and all the methods of designing like ladders, twists, accordion, and raindrops, as well as help to reveal the design further.

Creating mosaic Damascus is the next stage of intricate design welding. Although the actual components of the mosaic design are still undefined, yet the designs in mosaic Damascus can be seen on the edges of the bar. In 1999, while at the BLADE convention, I asked a few of the most proficient Damascus makers present about their viewpoint on this subject and they all defined mosaic Damascus differently. As the definition is still unclear, every end-grain design will be called mosaic.

Creating Basketweaves, Spider Webs, and Radial "W's"

To some extent, the parquet or basket weave is a simplified mosaic Damascus design and is suitable for a beginner to practice with. To start, hammer a low billet with five to nine layers and lengthen it into square bars one-inch in size. Divide the bar into four sections and pile them up in a square of size two-inch by two-inch with the horizontal layer in the opposite corners and the vertical layer placed in the remaining corners.

Hammer the pile and lengthen the bar out, maintain the square shape of the bar by hammering equally on every side Repeat cutting and welding as explained above creates a fancy basket weave design, a pattern that is suitable as a background filler in a complex mosaic project.

Another example of a basic mosaic pattern is the spider web also known as the spider grid.

To produce it, begin with squares if solid steel, like 1050 or 1095. Divide the steel into four squares and pile them up into a square billet, then shims of contrasting steel like 15N20 or pure nickel are added and then welded with a hammer. After this, reshape the billet into a square bar one inch in size. Divide the bar into four sections, pile up and weld again till the preferred grid size is gotten. The grid can be deliberately contorted by hammering it diagonally to give the design a special effect resembling a spider web.

The radial design is another type of mosaic Damascus design A low-layered billet of the flattened sheet is used to start the radial design. A die is used to divide the bar, and this causes an indentation in the center of the layers. Then the halves are divided into four sections, piled up into a square and welded together with a hammer; this creates an effect that makes it look as if the layers radiate outwards from the center of the square.

Exceptional design is created by applying the radical method to a "W's" design. For all the patterning methods, any preferred billet can be used; you might create a remarkable design.

Four-Way and Nine-Way Forging

To achieve the best result in the completed blade, it is necessary to include more than one radial or one jelly roll in the design. A bar can be sectioned into four, piled into a square billet where two sections are arranged in two rows and welded with a hammer; this is called a 'Four-Way.' Also, the bar can be divided into nine sections, and three sections are piled into three rows or a 'Nine-Way.'

To achieve the desired result, the 'Three-Way' or 'Nine-Way' billet can be replicated many times. The size of the blade to be created determines the number of original elements in the completed bar. For large fixed-blade knives, use a minimum of 16 original elements and do the 'Four-Way' two times to achieve that. For small fixed-blade knives, make use of 36 or 64 of the main elements in the bar. A 'Nine-Way' and a 'Three-Way' is needed to accomplish 36 elements whereas doing a 'Four-Way' three times achieves a billet of 64. These values are merely suggestions because the way a bladesmith utilizes the materials is determined by individual preferences.

A single design or several different designs can be mixed in various Four-Way or Nine-Way sequences that accomplishes fascinating designs that have high contrast. The possibilities for producing designs with these combinations are limitless. After combining and welding the preferred amount of elements with a hammer, there are many ways to reveal the design at the end of the bar. Bending the bar and then hammering it into shape will reveal the design along the blade edges. Like in any bent bar, the edge has to be moderately thick to enable more grinding because it is better for the design to be situated close to the center.

To reveal the design, the bar can also be hammered to a rectangular shape by compressing the ladder on the bar or grinding them in. Although I have never applied the raindrop technique, it should expose the pattern on the external part of the rectangular bar as effectively as the ladder method.

The Accordion Technique
This is my most preferred method of revealing an end-grain or mosaic Damascus design. You'll like the illusion of movements and flux that is produced using the Accordion method. Many different techniques can be used to expose a bar like an accordion.

The Damascus bar is hammered into the final

length and then hardened. A band saw is used to divide the bar to remove triangles of material from the opposite sides of the bar.

Once all the cutting is completed, a grinder is used to make the sharp edges rounder. This is followed by flattening of the bar. Optimum welding temperature should be used to work on the bar when the accordion is being flattened. If there is a rip in the bar at the bottom, use soldering flux and carefully weld it up. Make the entire bar flat by using one high temperature. Then the bar is hammered to its final length.

The cut-out triangle variant of the Accordion Method will achieve perfect results for you so you'll not need to employ other accordion techniques. The method requires extra hours of labor; but at this point, the Damascus bar is important, and you'll not be bothered by the effort that would only help you to make the most of the material gotten from the bar.

The Loaf Technique
The Loaf Technique is another method of revealing end-grain designs. It is achieved by welding several blocks alongside each other with a hammer. It is important to enclose the blocks with a protective material, like plain

carbon steel or Damascus. The seams are then closed up with a hammer, and the billet welded dry. The weld is made more simple if the blocks fit together perfectly. The Loaf Technique is effectual for designs or figures that require no contortion.

Another way to reveal the design details and create blade material is by cutting tiles from the bar, then using a dovetail to fit them tightly and welding them with a hammer. This technique is complicated, and a beginner is not advised to try this yet. Using a tack, the designs are welded to a protective plate that is ground off after welding with a hammer. This technique also doesn't contort the initial design.

The Plug Method
Now to the final technique used to reveal end-grain designs. The initial bar can be hammered or changed into a rounded bar and plugs are then clipped off. Drill a hole into a blade and insert the plug into the hole. Try to achieve a nice, snug fit while making sure that the plug is a bit thicker than the blade. The blade and plug is assembled and then heated to an optimum temperature and welded in hammer combination or one press. If you like, you can weld a lot of plugs into one blade; this

technique also doesn't create contortions.

Sometimes contortions can be used to improve a design or produce an entirely different design. The design in the bar is contorted by hammering a square be in a right-angled bias. The contortion remains as the bar is hammered on the bias until it becomes square back. This can then be used like that or combined into a Four-Way or Nine-Way.

The 'Persian Ribbon'

A design that is ordinary looking can be made more exceptional by contortion. After hammering on a right-angle bias, the squares in a Four-Way are transformed into triangles and can be modified again by another Four-Way to produce diamonds on the design; this method is called the Persian Ribbon. Four blocks are piled up in a square and birders of contrasting materials are wedged between the blocks. They are then welded with a hammer and rotated on a bias, the borders create an 'X' along the bar. The bar is opened with the Accordion Technique. This the 'X' produces the Persian Ribbon design.

Making Custom Images

After the mosaic Damascus pattern has been created, another step is the production of images inside the Damascus steel. It is not uncommon to place images in Damascus; I have seen blades with shamrocks, dragons, mammoth outlines, scenes showing bird-hunting, and many other objects. Utilizing finely ground steel has made it easier to produce these images and figures.

Before the advent of ground steel, an EDM (Electrical Discharge Machine) can be used to carve out an image from two blocks of contrasting steel and the male parts were exchanged. Welding with a hammer produced two bars that had the same figure, one darker than the other. This technique was highly-priced and tough complex detailings could be carved, hammering unevenly still created distortions. If ground steel is used, one block can be carved using the EDM and the figure is then extricated; followed by the pouring of the contrasted ground steel into the empty hole.

Put the male part a square tube and seal the end with a cap, contrasting ground steel is then poured into it and it is welded with a hammer; this produces two bars with the same shoes at half the price of using the EDM.

Another technique for producing images or shapes in steel is to cut plates and pile up the plates. Instead of using the EDM that is more expensive, laser or waterjet can be used to cut the plates. The plates are piled up in a square, and the contrasting ground steel is poured into the already carved out shape; then it is welded with a hammer. It is advisable that the ends of the pile are welded first. This is to compact the plates tightly and prevent powder from passing through the plates. While positioning the shape at the end of the square bar, the billet is then hammered onto the bar; this technique is cheaper than the EDM technique and the materials used are easily accessible.

Casting pure nickel around carved-out molds can also be used to produce shapes. I have carved out wooden shoes of shamrocks, birds, fish, and other objects. This is cheap and requires no tools like the EDM and laser. The nickel form is put in a square tube, and the tube is filled with ground steel. Different shapes can be hammered and put in the billet; you can place any object you want in the billet to achieve the desired result. If ground steel is used, you must compress the powder thoroughly before sealing the tube.

Allowing the tube to vibrate enables the powder to settle and become as compact as possible. While welding with a hammer, initially the bullet feels delicate, but it starts to harden after the bullet is decreased to one-third of its size. Some ground steel moves at a different speed while being forged, so it is important to have a keen knowledge of hammering the bullet with just a little contortion. Start with a basic project and concentrate on how things move in the bullet and the effects can be seen after a little while.

The use of ground steel only started recently; it was introduced in the early 1990s by Steve Schwarzer. There have, however, been a significant advancement in the Damascus steel sector. From my viewpoint, the uses of Damascus steel are endless, and we have barely started maximizing its functions.

Etching the Damascus Steel

There are different ways of engraving in Damascus steel. These are the guidelines on how to etch (or engrave) Damascus steel:

Step 1: Tools and Materials Needed

- Damascus steel

- Engraver-Ferric calories (strength- 42 name)

- Distilled water

- Wire (to suspend the steel in acid)

- Acetone

- Baking soda

- Oil (I used olive oil since the knife would be used for cooking)

- Rubber gloves

- Bucket of water

- Extra fine (000) steel wool

- Plastic or glass containers

Step 2: Steel Cleaning

- Ensure your steel is extremely clean.

- After cleaning, wear the rubber gloves and avoid touching the steel with your bare hands until the entire procedure is finished.

- Immerse in acetone and leave to dry.

Step 3: Acid Bath

- Mix one portion of ferric chloride with three portions of distilled water, and pour in a suitable container. Label the containers, so you won't forget what it is later.

- Hang the piece in the acid. Take it out after ten minutes and immerse it quickly in the bucket of water.

Step 4: Scrub with Steel Wool

- Use steel wool roving to scrub it until you have removed the oxides. The designs should look beautiful.

- Immerse in the bucket of water.

- Is the design intense enough? Otherwise, immerse it in the acetone (see the beautiful picture) and repeat the previous step until the desired result is achieved.

- The knife in the picture was dipped for ten minutes, three times.

Step 5: Repeated Process

- If the intensity of the engraving is to your satisfaction and you have scrubbed for the last time, immerse it in the acetone. Leave it to dry and to get the completed finish, dip it into the acid for three months.

- Avoid touching it at this point. Dip it in

the acetone again and then leave it to dry. Acetone dries up quickly so suspending it up would be very unnecessary.

Step 6: Bake and Shake

- Scoop a tablespoon of baking soda into a saucepan of water and heat the piece for ten minutes. This helps to neutralize the acid on the pieces and allows the oxides to set

Step 7: Oil, Not Spoil

- At this stage, use a 2,000 grit sandpaper to sand the steel gently to make it polished and shiny.

- Oil it immediately once you are through; if you don't, it will start to rust.

- If you want, you could apply a clear coat varnish.

Chapter 2. A Comprehensive Guideline to Japanese Forging

The Japanese technique of swordsmithing requires a great expenditure of effort. It was developed in Japan to create historical Japanese weapons (nihonto) like tanto, nodachi, tachi, odachi, naginata, katana, wakizashi, ya (arrow), uchigatana, nagamaki, and yari. Japanese blades were crafted in various shapes and thickness with different quantities of grind. Wakizashi and tanto are not just scaled-down Katanas; they were crafted with the absence of a ridge (hira-zukuri), or other shapes which were not commonly done on katana.

The art of crafting a Japanese sword blade generally takes days or weeks and was held at 'sacred,' followed by a great exhibition of traditionally religious Shinto rites. Like other couples projects, many craftsmen were involved instead of a single artist. One swordsmith hammered to create a rough shape, and another swordsmith (the apprentice) bent the metal; there was a polishing specialist, an edge specialist. Sometimes, some swordsmiths specialized in the hilt, handguard, or sheath.

Materials and tools

- File
- Vice
- Forge
- Tongs
- Anvil
- Hacksaw
- Bucket of water
- Square hammer, ~3 lbs
- Spray bottle with water
- Cross peen hammer, ~2-3 lbs (optional)
- Safety equipment (apron, gloves, safety goggles, etc)
- 7"×1/8"×5/4" piece of shallow hardening steel

Step 1: Forging

The content of carbon in the steel bloom or kera (manufactured in the tatara) varies; from pig iron to wrought iron. Three kinds of steel are used for the blade—the kind of steel used for the blade core (shingane) had low carbon content and was called hocho-tetsu. The outer skin of the blade (kawagane) was formed from a combination of steel with high carbon content, tamahgene, and the remelted pig iron (nabe-gane or cast iron).

Procedure:

The folding of the steel is the most understood stage in the production process. Here, the sword blades are created by repeated heating, hammering, and folding of the metal. Folding makes the metal stronger and free of impurities; this was an innovation of legendary Japanese swordsmiths.

1. Japanese swords are crafted by folding the low-carbon iron on itself many times to remove impurities; this yields the soft metal that is used for the blade core.
2. The high-carbon steel and the cast iron (with a higher carbon content) are then hammered in rotating layers.
3. The cast iron is heated, dipped in water, and divided into small sections to remove slag.
4. The steel is then hammered into a single plate while the sectioned cast iron are stacked on top and the entirety is welded with a hammer into a single billet; this is known as the age-kitae technique.
5. The billet is then lengthened, divided, folded, and welded with a hammer again.

The ways of folding the steel can be: longitudinal (sideways) or transverse (from the front backward). Both folding directions can be employed to achieve the preferred grain design; the procedure is called shita-kitae and can be repeated eight times or even up to 16 times. After folding 20 times (2^{20} or 1,048,576 separate layers), the carbon content would have spread evenly, creating homogenized steel; thus, folding no longer produces an effect. The quantity of carbon used determines the type of steel produced, maybe the hard steel used for the edge (hagane), or the moderately hard spring steel (kawagane) used for the back and sides.

In the last few folding, the steel may be hammered into many thin plates, piled up, and welded with a hammer to form a brick. The steel grain is delicately placed between close layers. Note that the portion of the steel used to produce the blade determines the actual arrangement

With each folding and heating, a mixture of water, clay, and straw ash are used to coat the steel to limit it from oxidizing and carburizing. The clay is to give room for a favorable environment. The formation of a wustite layer is boosted by the clay and water at about

1.650F, which is equivalent to 900C. This kind of formation occurs when oxygen is absent. The silicon which is already present in the clay will then come in contact with the wustite to form fayalite, and the fayalite will turn liquid at about 2,190F. The liquid we have now will stands as a flux, bringing and drawing out impurities when it's squeezed from the layers. With this, a pure surface is obtained which will then help the forge-welding process to be accomplished. With the loss of impurities, slag, and even iron as sparks during the hammering process, the weight of the steel will be reduced and becomes even lighter. This process is so popular now because it uses the impure metals which come from the low temperature that is needed in the smelting of it at this particular time and place.

Folding helps to achieve the following:

- It gives an alternative layer of different hardenability. This is to say that during quenching, the high-carbon layers becomes harder than when the medium carbon is used. These two the hardness of high carbon and low-carbon steels combine to form property of toughness.

- It removes any emptiness in the metal.

- It helps to hold the metal within the layers and making an even spread of other elements like the carbon throughout the individual layers which help increase the effective strength and thereby reduce the number of initial weak points.

- Folding helps to eliminate plenty of impurities which help fight the poor quality of raw Japanese steel.

- It helps to provide about 65,000 layers through daily decarburizing its surface and also bringing this too to the blades interior and also assisting in giving the swords their needed grain.

Step 2: Assembly

The high-quality Japanese sword is made from different distinct types of steel. The manufacturing process employs the use of various steel types at different parts of the sword. Maru (or muku) is the main type of modern wakizashi and katana, which is the most used, and the whole sword is made up of single steel. The good thing now is that the modern steels are being used which made the sword to be strong and no longer fragile like what has been registered from history. Another type—the kobuse type—is composed of two sheets of steel, which are called hagane, otherwise known as edge steel, and shingane, also known as core steel. The last types are called Honsanmai, shihozume, and kawagane.

There are numerous ways these steel can be assembled, which differs from one goldsmith to another.

1. The edge-steel is dragged out, bent into a U-shape and the softcore steel is then pushed into the hard piece of it.
2. This is then welded together and nailed into the basic shape of a sword.
3. At the end of this, the two plates of steel are put together.

The difference between the two is just in their hardenability. When it comes to more complex construction, it can mainly be used with antique weapons. Various modern weapons are constructed this way with either one, two, or three sections.

Another process that can be used is

1. Put together the pieces into a block and forge weld it together.

2. Drag out the steel and make into a sword so that the steels will end up being in its correct place as it ought to be.

This method is mostly used for more sophisticated models, which enable effective parrying devoid of fear of destroying the side of the blade. Now to make honsanmai or the shihozume types, it all about gathering the pieces of hard steel and putting it to the outside of the blade in the same way. It just that these types shihozume and soshu are not easily accessed, but they give very rear support.

Step 3: Heat Treatment for Japanese Blade

This is the point the Japanese process of sword making will be appreciated. When forging has been done, the steel will still retain its hardness. Without further treatment, this hardened blade will be brittle. Therefore, it is subjected to heating till it attains the right softness level.

Heat treatment is the tempering and quenching of the steel. Tempering is done by heating the blade to about 400°F, while quenching is done by rapid cooling.

Heat treating makes steels flex and strength to react in various ways. When steel cools quickly, it will be martensite, that is being very hard and brittle, but slower makes it pearlite which makes it bends easily and it doesn't hold an edge. But if one needs the combination of the two, a different hard-treatment should be used. In differential hardening or what is known as differential, quenching is a process whereby the sword is painting with layers of clay before heating and a thin layer is provided at the edge of the sword which boosts fast cooling to bring fast hardness for the edge. Afterward, a thicker layer of clay will be applied to it, too; this will bring about slower cooling. This will bring about softer, resilient steel, giving the blade the ability to absorb shock without breaking. Another name for this process is called the differential tempering, though this is a different type of heat treatment.

If one needs to bring out a difference in hardness, then the steel should be cooled at a different duration and again by controlling the thickness of the insulating layer. Even the Japanese by being able to control the different rates of the heating and cooling speeds of the sword blades, they were able to manufacture blades that had a softer body and a hard edge.

This Japanese process has brought about two major effects on the swords:

- Making the blade always to curve.

- Bringing a clear boundary between soft and hard steel.

When the heat is quenched, the less-heated edge will contract, which will make the sword to bend firstly toward the edge; but unfortunately, the edge can't contract fully before reaching the martensite forms, since the remaining part of the swords is still very hot and also in a thermally expanded state due to the heat. However, the sword spine will still be hot and pliable for several seconds, but will later contract even more than the edge which makes the sword to still bend away from the edge, of course, this helps the swordsmith have an easily curving blade. Again, this will make the differentiated hardness and the methods of smoothing the steel result in what is called Hamon, which means hardening pattern or tempering line. The Hamon can be described as the noticeable outline of the yakiba, which is used as a means to assess both the beauty and quality of the finished blade. Hamon patterns come from the various manners in which the clay is used. There can act as some style for sword molding or better still a mark that a swordsmith is known for. The difference in how hard steel can be done better when put near the hamon. This will help to show different layers or parts of the blade which includes the intersection between an edge from the edge-steel and the sides which are produced from the skin-steel.

To get the right thickness of the edge's coating then there should be a right balance of temperature of water. Also, the right hardness of the edge can be achieved without one needing any tempering. In some cases though, the edge will turn out very hard, so the best bet will be equitably tempering the whole blade for a while so that the hardness will be to a minimal level which will be suitable for the edge. The suitable hardness normally is between HRc58-60. Heat treatment helps reduce the martensite hardness and make it become tempered martensite. Pearlite doesn't work well with tempering and doesn't even change in hardness

Step 4: Finishing (Polishing, Mounting, and Sheathing)

When the swordsmith fully does the rough blade, the smith will then put the blade in a polisher, which is known as togishi. The work of togishi is to sharpen the shape of the blade and add artistic value to the blade. This whole process takes ample time and might run into weeks. Olden days polishers were using like three types of stone, but these present polishers are using seven types. The modern high level of polish came to be around 1600 because then the emphasis was placed more on functionality than its form. Polishing processes always took more time than crafting, which helps to heighten the beauty of a blade is usually good polish while bad polisher damage even the best of blades. A novice in the polishing business will create a bad blade which can forever damage it.

Saya in Japan is known as a scabbard for a katana. The handguard piece is created as one's work of art, and in the Edo period, it was known as tsuba. Mountings are done when the blade is finished, and the job is passed to the mountings-maker also known as a maker. Sword mountings take different shapes in a different era but wholly, it all about the same idea, the difference only lies in the elements used and then the form of wrapping. The visible part of the hilt includes a metal or a wooden grip, which is known as the tsuka and can be said to be the entire hilt. The handguard or tsuba which is on the Japanese swords which are very small and round is produced of metal and often stylish. It has a wrapping called menuki, and there always a decoration beneath it, and it has a pommel at the base which is called a kashira. A bamboo peg known as mekugi is fastened around the tsuka and also through the tang of the blade, the drilling of it is done with a peg hole. This helps to set the blade securely into the hilt. To fasten the blade tightly into its sheath, the blade has a collar or what is known as habaki which stretches an inch past the handguard, and this will help keep the blade from damages.

Producing sheaths aren't easy; we have two types of sheaths, and the two require great work to produce. One type is the shirasaya, which is largely created from wood and is known as the resting sheath, which is used as the storing sheath. The other type is the decorative and battle-ready sheath, which is also known as a jindachi-zukuri, that's used from the obi, by straps or buke zukuri if it thrust through the obi. Kyu-gunto, shin-gunto, kai-gunto are all types of mounting that were used in the 12th-century military.

Chapter 3. Stainless Steel Forging of Full Tang Knife

This is about how a full tang knife is made from stainless, though it is good to do self-research because the climatic conditions also affect the stainless steel. Creating one's tools has always been an awesome thing, knives inclusive. From the stone age until now, knives have always been a valuable part of a man either as outdoor protection or what should be used in the home as a basic tool and, of course, it is what should be produced to your taste and style. Well, we have numerous ways on how we can produce a knife. There are tried and tested methods, but this underneath method can work excellently for stainless steel forging.

Materials/ Tools needed:

- Files
- Drill
- Dremel
- Epoxy
- Forge
- Anvil
- Forge
- Clamps
- Grinder

- Hammer
- Tongs (or big pliers)
- Hot cutter
- Quench tub
- Linseed oil
- Angle grinder or hacksaw
- Steel (5/32" X 1 1/4" X 96")
- Handle Material (see handle section)
- Sandpapers of varying grits from about 120-1000.

Step 1. Designing the knife on paper— Materials needed

Graph paper, Metal Ruler, French curve, Normal pencil

Generally, the best bet in making anything is first to design it and doing the same with a knife isn't a bad idea. Having the material display of one will enable you to see all that you need to work with.

Step 2: Basic design on paper—The blade and the handle

The main focus when it comes to knife production should be the blade, and the handle and your focus should be on this two; nonetheless, time can still be put in other parts like the pommels and the guards. If you have a band saw then other parts would be easy to make. Freehand rendering can be used for the designing of the knife. If you like the basic proportion, then you can go ahead. It is always good for you to design what you like and what looks delightful to your eyes.

Step 3: Basic design on paper— Designing the blade

The first thing a person notices about a knife is its blade and as such more attention should be paid when designing it. The blade should give the knife a better outlook and uniqueness. You can make do with a simple blade with nothing complex about it. Starting with something simple will bring about less grinding, and it will perform better. So you should start with cutting about an 8-inch by 1.75-inch blade, then drew out a curve, design it to taste and get it ready to be taped to its stock.

Step 4: Making the blade proper— Melting and casting

First, melt the raw materials together. This will be done in an electric furnace. Usually, this step requires intense heat of about 8 – 12 hours. After melting, cast the molten steel into semi-finished forms, including slabs, tube rounds, rods, blooms of rectangular shapes, and billets of round or square shapes (about 1.5 inches thick).

Step 5: Making the blade proper— Forming

After melting and casting, the semi-finished steel undergoes its forming and shaping operations. This starts with hot rolling, whereby the semi-finished is heated and passed through huge rolls. The billets are formed into a short bar of ¼ inch or ⅛-inch thick (for a full-tang knife).

Step 6: Making the blade proper—Heat treatment

Heat treatment (also called annealing or tempering) is done after the steel has been formed and shaped. This is the heating and cooling of the steel under specific conditions to make the metal soft. Whether it's austenitic, ferritic, or martensitic, different steels have different temperature requirements. Austenitic steels are heated at temperatures above 1900°F, depending on its thickness. It is sometimes important to clean contaminants from the surface before heat treating.

Some steels also go through age hardening, a heat treatment for higher strength which is done with great care. The properties of the steel can be affected by changes in recommended temperature, cooling rate, or time. Higher strength is produced from lower aging temperatures while a lower strength and tougher material is formed from high-temperature aging. The heating rate to attain the required aging temperature is 900 to 10000F.

Quenching is the rapid cooling done after heat treatment. It increases the steel's toughness while retaining its strength. One recommended process of quenching is by using water, whereby the material is dipped in a 35°F ice-water bath for at least two hours. Thick sections are quenched using water while thin sections are quenched using air cooling or blasting. Cooling must not be done slowly so as to avoid carbide precipitation. Thermal stabilization can be done to eliminate buildup by holding the steel at 1500 – 1600°F.

Step 7: Making the blade proper—Descaling

After tempering, scales or build-up are sometimes formed on the steel. These can be removed in several ways—one of which includes pickling. Pickling is done by bathing the steel in nitric-hydrofluoric acid to descale it. Another method of removing the scales is electrocleaning. In electrocleaning, a cathode and phosphoric acid are used to apply electric current to the steel's surface to remove the scale. Depending on the steel type being used, tempering and descaling of the steel are done at different stages.

The knife steel bar goes through added forming steps including more hot rolling, forging, or extruding after the first hot rolling before annealing and descaling are done.

Cold rolling is the passing of the steel through rolls low temperature. Cold rolling leads to reduced thickness and makes the steel ready for final processing.

Step 8: Making and grinding the shape of the blade

This is one step that doesn't need you exerting much energy considering it a simple step; the only tool needed here is the bench grinder. Though you can still use an angle grinder, the con is that you can hardly see what you are doing and even getting straight edges might be difficult. Ensure you have your safety gear on and wear clothes that you might not be bothered with any longer after the grinding since you would be covered in metal. In this step, the knife will be taking shape already.

Step 9: Making the blade and grinding the Bevel

You have to be careful with this step because the cutting edge is like the most important process, so when trying to set it, you need to go gently and slow with it. So on each side, try to grind away half of the edge and be sure that you tightly clamp the blank. Maybe a C clamp can be used since the cover less of the knife considering it wouldn't allow it to move anywhere while grinding. To form the bevel use a 36 grit wheel for the angle grinder because it will help create a wide cutting path. When forming always look down the blank to check the angle, when you're halfway gone, flip it over to the other side and grind. Watch out when once you see a wire edge forming then you can stop, because you have succeeded in making a perfect edge.

Step 10: Making the blade—Rust removal and finishing

This is the part where the almost finished knife is taking shape. Finishing is all about making the knife shine; it is another important process in producing a knife. To achieve this use an angle grinder with a grit flap wheel of 120. Remember to clamp it tightly to the Adirondack chair, concentrate your effort on the blade, and maybe on the handle area too, so clean them to remove all the rust to look very clean, then hand sand it by clamping the knife tightly to a block of wood then clamp that in a vise. Start with 180 grit and work it up to about 1,000 grit. This will help give the knife a nice smooth satin finishing. If you still desire, use rough sandpaper on the handle to help the epoxy stick more. Stop when you have a nice shiny blade and move to the next step.

Step 11: Making the handle—Rough cutting and gluing

To begin this step, you need to start with creating the handle scales. Scales are those pieces that we see on either side of a blade. To make the scales, the size should be measured out by putting the knife on the wood and marking out the largest dimensions that are needed. With a square, mark out a rough rectangle, and cut it off, this can be done with the aid of a table saw since it's faster and easy to use, but hand saw is still good though it needs a lot of patience. Cut out the scales and then glue them together, mix up a good bit of glue and spread it on the scale and then place the knife on the scale, repeat it also on the other scale. Get them lined up and clamp them, you can wait for some hours like two hours before using the knife; this will give the knife plenty of time to set.

Step 12: Creating the Handle—Getting the Right Thickness

My findings have shown that knife handles should not be thicker than ¾-inch, but if the knife is bigger than the usual, you'd have to increase the thickness a bit. Adjust your calipers to half of your desired thickness and begin to scrape with a rasp. A plane can be used if you like. Keep filing till you achieve the right thickness on both ends. If the rasp you are using is sharp, this shouldn't take time; ensure to file equally on both edges.

Step 13: Creating the Handle—Rough Shaping

This is the most tiresome and lengthiest step. Here, the handle closely resembles a handle and not a block of wood. You need to be insightful for this step because you are required to know the point where the metal starts. Striking a wood rasp on metal makes it duller. The simple logic is to continue using the rasp until you can feel that you are getting close to metal, and then change to a file that can be used for both wood and metal. This part is moderately tiresome, but you can't afford to rush when doing this. You can now see the primary design taking form which is what is required at this stage. Also, you might want to wrap the blade using duct tape so it can be clamped in the vise and you can keep working on the entire handle.

Step 14: Creating the Handle—Final Sanding and Fixing Pins

You are almost through, all that is left is sanding and pins. As the name implies, sanding is working through the grits till it is about 320 or 400; you can play around and increase or up to 1,000, but I am not sure it would change the texture of the knife. The knife handle here was sanded up to 320 grits before final touches were added.

I would be explaining how to fix the pin and adding final touches since they are done at the same time. Drill a hole through the handle and blade, glue the pins in with epoxy and then file them flat.

Marking the holes is the initial step in this process. Make a rough estimate of the marks first, then check if they are aligned using a straightedge. The next step is to measure your pins and cut them. To measure, place them on the handle and use a sharpie to mark. Create room for error; make it a bit bigger. Removing material is easier than adding to it. Using a vise, clamp the rod and use a simple hacksaw to cut the pins off.

Next is the drill press. Clamp the piece firmly and drill through everything at the same time. It's best to set the drill to the lowest speed and use a very sharp bit. Do this again with as many pins as you desire. When you are through, recheck how well the pins fit in the holes. You are very fortunate if they barely fit; tap them in using a mallet and file the ends off. If they are loose instead, then it's time for epoxy glue! Make another epoxy mixture and rub a little at the middle of the pin, slide it in and clean off left-over epoxy with a dry cloth. Before filing, allow to dry. When you are sure it is dry, use a mill file to file the pins.

Step 15: Make the Handle—Complete it all!

This is the final production step! This step is not compulsory. You can decide to have your handle left as bare wood if you like. You could also treat it with a little Danish oil and two coats of polyurethane. The method for applying oil is easy. Clean off the handle with a clean cloth, then using a rag, apply oil; ensure to cover the entire handle. Used two coats and left them throughout the night. Before adding the polyurethane, with the finest grit earlier used, mildly sand the surface using 320 grit. Then make sure to clean off all the sawdust again with a clean cloth. Apply a very light coat of polyurethane to the wood using a film brush. Use two coats and sand before each coat. After the last coat, leave it for 24 hours. This allows it to set fully. Congrats! The manufacturing process is complete.

Step 16: It is time to sharpen!

Your knife finally becomes a tool at this stage. It is important to get a razor-sharp edge on a knife, but this can be a bit hard with a handcrafted knife. Using your hand to grind the edges unavoidably results in irregularities, but this is expected as part of handcrafting your tool. Little errors like this can be fixed, so do not despair. To sharpen the knife, it's best to use a sequence of stones: a coarse, followed by medium, and then fine stone. Work through your stones till you achieve a very sharp edge. After this, you can choose to leave it that way or sharpen the edges on a strop. This gives your knife a razor-sharp edge. To try out the edges, slice up some paper, it should cut through perfectly. If it doesn't, continue. When you can constantly slice up paper into ribbons using your knife—you are through!

Step 17: Congratulations!

At the end of this, your full tang stainless steel knife is completed. There is a feeling of contentment when you use a tool you made by yourself and see it work equally as good as, or better than knives bought in the store. Additionally, it is a good skill to be able to create a tool from scratch (in case the zombies arrive!).

Be careful and enjoy your new knife.

Chapter 4. Exotic Handle Making, Wood, and Metal Finishing

Knife handles and handcrafted swords have a distinct attractiveness to them.

Sword handles are fashioned to support and improve the beauty and safety of a specific sword type. The sort of handle you would create must properly for your selected sword. The procedure for creating custom-made knife handles may take a lot of time, but it is simple as long as you know the steps.

Luckily, you would eventually have a beautifully customized knife to display proudly. This segment will help you learn how to make the most common European sword—a hidden tang, screw-on pommel sword handle. Because handcrafted items do not have generalized specifications, measurements are determined by the blade you are designing the handle for.

Materials/ Tools needed:

- Files
- Scribe

- Clamps
- Epoxy
- Drill Press
- Knife Blade
- Metal Sanders
- Metal Band Saw
- Wood Band Saw
- Masking Tape
- Sand Papers (200-600)
- Handle material (could be wood or plastic or stone etc)
- Pins - Mosaic, Solid, or Rivets (optional)
- Color spacers to add lines of color between metal knife handle and your wood (optional)

Steps to Creating Knife Handles

Part 1. Getting the Base and Materials Ready

1. Create or get a full tang knife blade—the constituents of a full-tang knife blade are a bare metal handle and a metal blade. The part of the handle (tang) should already be cut in the same shape as the wooden handle (scale). You can shop online for knife kits that have only the full-tang blade and pins.

2. Cover the blade end of your knife with

three layers of tape.

Duct tape or electrical tape would be perfect. Masking tape can be used, but more layers would have to be added. Cover the tips of the knife through to the base—the point where the blade ends—with the tape. Do not wrap the tang. Ensure that the blade is entirely covered with the tape; this will prevent the knife from cutting you, or epoxy glue from dropping on the blade.

In addition to preventing you from getting cut, the tape also protects the blade from breaking off or scratching.

If you can still touch the blade through the tape, cover the blade with more layers of tape.

3. Select two 1/4-in (0.64 cm) pieces of wood for the scales. Select a strong, long-lasting wood about 1/4-in (0.64cm) in thickness and a bit bigger than the tang. For a better finish, make sure the grain is of the same length as the wood. These pieces can be bought online from stores that deal in knife-making materials.

A knife handle is made of two halves which are also called 'scales.' The tang is placed in between the scales.

Excellent examples of wood to use are pear, apple, ash, hickory, bois d'arc, and pecan. Hardwoods are gotten from deciduous trees and generally last longer than softwoods gotten from coniferous trees.

4. If required, cut the ponds. If you bought a knife kit, the pins might have been cut already. If you didn't buy a kit, you are required to cut the metal rod onto one-inch (2.5 cm) lengths. Place the rod on a stable surface and use a file or a metal bar to cut it into one-inch (2.5 cm) lengths.

The number of holes in the tang determines how many rods can be cut. While a few knives have two holes, others have four. The metal rods have to be thin enough to fit the holes in the tang correctly. For every knife, the thickness of the rod differs.

5. If required, file down the pin ends. Again, if you bought the knife kit, the pins would be already filed; but if you

cut the pins by yourself, the ends will probably be bristled, smoothen it using a metal file or a grinder.

Do not be distressed of the ends of the pins are not completely flat. You would still file them down later to make them aligned with the scales.

6. Using a plastic wrap and plywood, line your clamps and vice. Though you would not need this until it is time to glue everything, it takes a short time for epoxy glue to set, so, it would be easier to have prepared everything. Connect plywood to each end of your vice. Get a sheet of plastic wrap, fold it like a taco and tuck it between the vices. If you have one, use heavy-duty vice mounted on a table. If you don't, instead, use two or three vices.

The plywood prevents the vices from making an indent on the wooden scales. The plastic wrap prevents the epoxy glue from dripping around. Wax paper can be used in place of a plastic wrap.

Part 2: Drilling the Pin Holes

1. Wrap duct tape around both the scale

and tang. Pile the scales on top of each other with your desired sides placed outside, with the handle facing out. Place the tang on top and wrap a piece of masking tape around the center to grip everything firmly.

Ensure to leave the holes in the tang open.

If the wrapping tape is not firm enough, wrap another layer of tape around the end of the scales and tang. Masking tape is perfect to be used here because it has a powerful hold and also leaves just a little residue.

2. With the tang's hole as a guide, use a drill press to make the first hole. With the tang facing upward, place the knife down on the drill plate. Put in the drill bit into one of the tang holes. Power up the drill and press down on it. Ensure to go through both scales. Power off the drill press and bring lift out the bit. With a drill press, this is much easier to do; a handheld drill can do the same thing, too.

3. Insert a pin in the tang hole and do the rest. If there are four holes in your tang

instead of two, the position of the second hole oblique is oblique to the first. Insert the pins and drill the remaining two holes obliquely across each other. Insert the pins immediately you are done drilling the holes. Drilling the holes and inserting the pins one after the other will prevent the tank and the scale from slipping.

Tap the pins in using a hammer.

4. Take off the tape and trace the tang onto the scales. Remove the piece of the tape, but do not touch the pins and tang. Using a marker, trace the outline of the tang. The type of marker used- whether temporary or permanent is not important, as you will still sand this off later.

5. Take away the tang and cut off the scales. Set the tang and do not take the pins out of the scales. Outside the outline you have traced, cut the scales with a band saw, or a scroll saw. Later, you would sand the scales to suit the tang.

You have to cut through both scales concurrently. The scales will be held firmly together by the pins.

6. Sand and shine the top edge of the scales. After you put the knife handle together, it would be impossible to sand and shine the narrow edge on top that is touching the blade at the base.

The blade might cause an obstruction, so you had better do this now. Just tape the scales together, sand and polish the top edge to your satisfaction. Shape the edge with a belt sander. Use a 220 and 400-grit sandpaper to sand down the edges. Complete with a buffer. It would even be better to get the pins inserted into the scales. This would make sure that the scales are well-positioned and equal.

Part 3: Gluing the scales

1. Clean both edges of the tang to get rid of oils and dirt. You could use rubbing alcohol or a window cleaner. Just rub your preferred solution down the tang and allow to dry. After this, do not touch the tang with your hands alone.

Rubbing alcohol gives a better result, but window cleaner can be used as well.

It is not necessary to clean the wood scales. The texture and porosity of the wood enable it to absorb the epoxy completely.

2. Scrape the tang on both ends to allow the epoxy to adhere firmly. A metal file or a screw can be used for this. No precision is needed for this step, but you have to clean the surface properly as soon as you are through. If there are no bumps or irregularities on the marked side of the scale, it would be good if you scrape them too. The scales can also be sanded roughly with 120-grit sandpaper.

    Ensure to scrape only the sides that would be in contact with the tang.

3. Follow the instructions and mix the epoxy glue. There are different brands of epoxy glue, but mostly, you will be required to prepare the same amounts of 'Quantity A' and 'Quantity B' in a disposable plastic cup. Do this quickly because it takes a short time for most epoxy glue to set. Confirm that you are using epoxy glue and not epoxy coating or resin.

Stir the mixture using a disposable tool, because the glue damages whatever you mix it with. Wearing some plastic or vinyl gloves are recommended. Epoxy glue can be purchased in hardware stores; some craft stores also sell them.

4. Using the already prepared epoxy, glue the first scale to the tang. Evenly spread a layer of the mixed epoxy, using a disposable knife or a paint spatula on one of the sides of the tang and the side marked on the matching scale. Press both the scale and the tang together.

5. Get the pins inserted and the second scale glued. Working very fast, turn the knife over so you will be able to see the tang from the other side. Get the pins inserted into the holes. Coat the tang and the marked side of the scale left, and then press them together.

   You might need to hammer the other scale to make sure it fits firmly. You can use epoxy to coat the pins if you like. This will further strengthen the bond.

6. Place the handle into the vise and firmly clamp it. Ensure that the handle is inserted between the pieces of plastic

wrap- in this manner, the epoxy glue would not spill around. Clamp the vise as firmly as possible.

7. Clean excess epoxy with an acetone-soaked rag. After pressing the two handles together tightly, excess epoxy drips down. With a rag dipped in epoxy, clean excess epoxy that might have dripped down from between the scales.

8. Leave the epoxy to set. The type of epoxy used determines how long this would take. Some are set and available for use under one hour; others could take up to a whole day to dry. Confirm the hours it takes to dry and instructions on the label.

Part 4. Completing the Handle

1. Remove the knife from the vice. As soon as the epoxy has set, loosen the vise and remove the knife. Still, leave the tape around the blade.

2. If required, grind away excess pins. Use a belt sander or grinder to grind off excess pins protruding out of the surface of the scale. They have to be aligned with the scale. Sculpt and shape the belt

handle with a belt sander.

3. Continue sanding the scales up till the metal part of the tang. Sand off any lines left from tracing the outline of the tang onto the scale. Now, you can sand off the edges of the handle to make them more rounded and easier to handle.

4. Sand and polish the scales. Start sanding the scales using 220-grit sandpaper. When the wood is smooth, switch to 400-grit sandpaper. Complete with a buffer until the sales are polished to your utmost satisfaction.

5. If you like, seal the handle. For a better finish, you can add two coats of oil-based polyurethane and one coat of dewaxed shellac. After it dries, polish it until it shines. The type of brand used determines the amount of time it takes to dry. So, make sure to read the label properly. The time taken could range from a few hours to some days.

6. Take off the tape from the blade. Your knife is now finished and ready for use. You can scrape off any epoxy remaining on the blade using a craft blade, but ensure to do this along the length of the

blade. Acetone can also be used to dissolve the excess glue.

Chapter 5. Exotic Hilts Making

The handle of the sword is called the hilt (sometimes referred to as haft). The hilt consists of a guard, pommel, and grip. The guard might contain a quillon or cross-guard; sometimes, though rarely, a ricasso might also be present. A sword knot or tassel might be connected to either the pommel or the guard.

Pommel

The origin of the word pommel is Anglo-Norman, and it means little apple. The pommel is an enlarged fixture at the top of the handle. Initially, they were created to stop the sword from sliding off the hand. In Europe, during the 11th century, they were designed to be weighted enough to provide a form of stability for the blade. This allowed the sword to be balanced close to the hilt, giving room for variability of fighting techniques. The opponents may also be struck using some (as in the Mordhau technique); this depends on the specific sword design and swordsmanship technique. Pommels have been shaped in different forms. Various designs can be etched onto them and are sometimes decorated and adorned with jewels. In the 1964 book, *The Sword in the Age of Chivalry,* written by Ewart Oakeshott, he established a classification system for medieval pommels, in addition to his previous blade typology. Ewart Oakeshott pommel types are represented with alphabets A to Z, while the variants are represented with numerals.

A - A derivative of the typical Viking sword, the classical "Brazil-nut" pommel.

B - A shorter and rounded form. B1 is the subtype with a straight lower edge- "mushroom or tea rosy."

C - The derivative of the Viking sword in the shape of a "cocked-hat."

D - A larger sized and later modification of C.

E - A modification of D with an angular top.

F - A modification of D with a much more angular top.

G - A plain disk. G1- disk pommel designed with flower-shaped ornaments. G2- disk-shaped variant with a shell-like ornament. Both are specifically made in Italy.

H - A disk with beveled edges. One of the most

widely used types, seen in the 10th to 15th century. H1- an oval variant of H.

I - A disk with wide beveled edges; though the internal disk is smaller in size than H.

J - Similar to me, but here a deep groove is made in the beveled edges. J1- a modified variant of the classic wheel-pommel.

K - A modified variant of J with extensively wide and flat edges, common in the Middle Ages.

L - "trefoil-shaped," tall, rare, seen exclusively in Spain in the 12th and 13th centuries.

M - A subsequent modification of the multi-lobed Viking pommel type, seen commonly on tomb figurines in Southern Scotland and Northern England around 1250 – 1350, rarely seen now as only a few are remaining. Check Cawood sword.

N - In the form of a boat, not commonly seen in art or even surviving specimens.

O - An uncommon variant formed like a crescent.

P - An uncommon variant shaped like a shield. The only known example is seen at Nuremberg cathedral, on a statue.

Q - Pommels shaped like a flower; only seen from artistic sword drawings.

R - Uncommon pommel in the shape of a sphere. Only existed in the 9th and 10th centuries.

S - An uncommon cube-shaped type with the edges removed.

T - Shaped like a fragrance stopper (looks like the stopper on the top of a fragrance), also looks like a fig or pear. First seen in the early 14th century, but became common after 1360 with subsequent modifications till the 16th century. T1 and T2 are the main subtypes.

U - Seen only in the late 15th century, shaped like a key.

V - Common in the 15th century, shaped like a 'fish-tail.'

W - In the form of a 'misshapen wheel.'

Z - In the form of a square. Variants can be used to pinpoint the area and era found. Z1 and Z2b variants were common in South Eastern Europe. Z3 was typically Venetian swords shaped like a 'cat-head.' Z4- common in Serbia and Bosnia.

Grip

The handle of the sword is the grip made from metal or in some cases, wood, normally covered with shagreen (tough, untanned leather or the skin of a shark). Sharkskin lasted longer in temperate climates but lost form in hot climates. Rubber later became widely used in the late 19th century. Although other sword types might use the skin of a ray fish, referred to as "same" in katana construction. Regardless of the material used to cover the grip, it was glued on and supported with a wire helically coiled around.

Guard

It has been commonly misinterpreted that protection is provided by the crossguard for the user's whole hand against the opponent's swinging. The full handguard was only used when the shield and armored gauntlet were abandoned. There are no actual guards on the early swords; they only had a stopper-kind of material to stop the hand from moving on to the blade when thrusting.

Beginning from the 11th century in Europe, the guards began to take on different shapes. They were shaped like a straight crossbar (later referred to as quillon) at right angles to the blade. And from the 16th century, they were made more sophisticated with branches, curved bars, and loops to give the hand protection. One curved piece side-by-side with the fingers (approximately parallel to the blade or the handle and at right angles to the cross-guard) was called 'knuckle-bow.' Subsequently, the bars could be complemented or substituted with metal plates that were pierced decoratively.

The word 'basket-hilt' was then coined to describe these designs, and there were different types of these basket-hilt swords. At the same time, the insistence of thrust attacks with small swords and rapiers exposed the level of unprotectedness of thrusting.

By the 17th century, there was a production of guards that introduced a solid shield that covered the blade externally up to the diameter of about two inches or greater. Preceding variants of these guards retained a single quillon or quillons. But quillons were absent in subsequent variants, thus, they were all together referred to as 'cup-hilt.' The guards of modern foils and epees were created using this subsequent variant as the prototype.

Ricasso

This is the dull section of the blade directly beneath the guard. It is protected by an extension of the guard. The ricasso allows the position of a third-hand on two-handed swords enabling the user's hands to be spread wider for a higher advantage.

Hilt Assemblage

To maximize the sword functions, it is important that the hilt is made the right way. The weight of the pommel must be adjusted to allow for efficient lifting and use of the blade. Guard and pommel have to be assembled with a strong and accurate fit. The grip must be shaped in such a way that permits good blade control and secure purchase. The hilt should be made with definite specifications fashioned in a style that is concordant with the era and type of the sword.

To be acceptable, the components of the hilt must reflect the subtle difference in the volume and shape as seen in original hand-crafted components. Wax models that show the dimensions and forms of hand-forged originals can be made to be used as a prototype to be borrowed by manufacturing companies.

Instead of handcrafting every original piece—which would be time-consuming and exorbitant, creating an investment casting by using a hand-crafted wax original can be done to achieve the same results.

It is recommended to use the higher priced and more accurate investment casting—also referred to as the 'lost-wax' method—instead of the less expensive and easier sand-casting procedure. Sand casting can be suitable for specific applications, but the intricate details required for the procedure I explain in this section cannot be produced using this technique. Producing the hilt component from a wax original to a complete piece ready for assemblage on a sharp blade requires a lot of efficient working hours.

Step 1:

A prototype for each component is sculpted from a block of solid modeling wax. This process requires extreme care and great effort; thus, sculpting and polishing are done for many hours.

The wax prototype must reflect all the detailing and show the distinctive shape of the prototype. The precise size of the pommel is achieved by testing the original blade for maximum balance in handling and performance. The quantity of the original wax is estimated, creating a space for the type of material used in casting the pommel (steel and bronze possess separate weights) and a contraction in the molding and casting process (roughly 3 – 5%).

Step 2:

The wax prototype must be shaped into two symmetrical halves as soon as it is made. The production or investment waxes will be created using this prototype. It is important that the mold is free of irregularities and long-lasting enough to produce wax each time so that the details would be exactly like the original and would need little to no finishing.

Step 3:

Before adjustments, rough investment waxes are made from production molds. Waxes are poured out into this mold according to a manufacturing schedule; demands for these pieces must already be prepared. Each wax is poured, left to cool and then extricated from the mold. Check the waxes for irregularities or flaws on the surface which are then fixed or condemned. Pouring, checking, and preparing waxes for casting all requires long hours.

Condemned waxes can be melted again to enable them to be used again. Every investment created from these molds is used once because the process of investment damages them.

Step 4:

The waxes are 'sprued' before investments; this implies that 'branches' of wax are connected to act as pouring vents or gates (to facilitate the exit of gases from the molten metal)

After 'spruing' the wax, it is dipped in liquid investment solution—a suspension that resembles ceramic (for more intricate pieces, a brush or spray painting is used to pre-paint investment solutions on the wax. The same treatment used for a ceramic or clap pot is also used for this coating—the coating is left to dry and fired in a kiln in an inverted manner to not only harden the ceramic shell but also to 'burn away' the investment wax.

Step 5:

Grind away the large sprues in the hilt components precisely. A first refinishing is done on the total component (because the investment constantly creates a slightly 'gravelly' surface that has to be polished off) and the pieces checked for irregularities. Other detailing that has been lost during the casting process can be regenerated carefully. Some small irregularities are often left deliberately to reproduce the 'character mark' seen on handcrafted hilt component. Following the restoration of all the details, prepare the hilt component for assemblage. Although a few processes like adjusting rough castings are done in sets, every sword assembly is treated individually, not only for craft but also for quality control purposes. The final assemblage is started by mounting one set of casting is on a particular blade.

Step 6:

The guard is brought very close to the base of the blade by hand filing. It is then hammered to fix it firmly on the shoulders of the blade. The hammer is then used to 'peen' the edges of the tang opening on the top of the guard to wedge the sides more tightly against the tang. Check the tight fit of the pommel on this NextGen Baron Next. The pommel is fixed to the end of the tang; this requires additional hand filing of the tang slot before hammering the pommel into position to make sure that it seats tightly in the correct position. After this, the pommel is then hammered to the end of the tang, wedging it firmly into place at the blade end. The end of the tang sticking out of the pommel is then filed into shape in preparation for peening. In some designs, an ornamental rivet block is included as an additional piece (most times, a truncated pyramid shape) that sits in position on top of the pommel.

Step 7:

Depending on the demands of the specific sword models, while cooling, the peen can be hammered flush or into an irregular decorative shape. The pin is then finished with care by being ground flush and polished, or filed into an ornamental block of peen. When a wet block is needed, the tang is 'peened' on top of the end of the wet block, and the peen and net-block are hand-filed to the finishing form.

Note: Because the pommel is permanently wedged into position on the tang, the peen isn't responsible for holding the sword in place. Each component sits without the support of the other.

Step 8:

A custom-made, stabilized birch core is then used to hand-fit the tang. Stabilized wood is used as it is resistant to expansion and contraction from humidity and weather changes and has greater strength and is more long-lasting for heavy use. The model of the sword determines the volume and shape of the core. Once the core is fitted, and epoxy glue is used to fix it into place permanently, risers and other details are added (as in the past) with a cotton cord or linen. The spacing, diameter, and length of cord riders are different and determined by the accuracy demands of that period.

Chapter 6. Leatherwork for Knives and Scabbard Making for Swords

How to Create a Sheath for a Knife

There are times when you would want an item to be as near as possible— exactly where you need it, not inside your pocket. A sheath can be created for a pocket watch, knife, cell phone, compass, or other objects you don't want to search for. You might like to create something to keep a multi-tool or any tool you want to be close every time. This section is beneficial for the knife you (may have) produced and it also helps you learn leather wet-shaping techniques. Leather can be expanded and molded in the shape of different objects once it is saturated.

Materials and Tools required:

- 5 – 6 ounces of medium-weight leather

- X-ACTO knife or Rotary cutter

- Needles to stitch the leather

- Saran wrap

- Dish towel

- Pencil

- Cardboard from a file folder

- Rowel wheel

- Fid

- Groover tool

- Thread or artificial sinew, already waxed

Steps for Making a Leather Sheath
Step 1. Pattern Drawing

Place your knife on your piece of cardboard; then outline the blade and the extent of the blade you want to be sheathed. The design is not equal because the back of the sheath has an elongation that will be folded eventually and stitched to create a loop that your belt would be threaded through. Note that perfection isn't necessary and it should be preferably oversize than too small.

Step 2. Make a Cut-Out and Assemble Your Pattern

With a pair of scissors, cut roughly to have an idea of the way your pattern would look when the knife is laid out. If this is to your satisfaction, fold the design in half, along the line that would later become the back part of the blade, and cut off the excess material so that the design is well-formed. Press the paper against the blade to know where it is located on the design. After doing this, what you see is a slight wrinkle on the cardboard that shows excess space between the blade and the edges of the cardboard. Then use a little adhesive tape to create the design in the exact three-dimensional shape your leather will be; this permits you to make alterations now when it is easier. Clip off the excess design to level it out and expose the handle a bit more. A few more trimmings and you'll be able to cut out the actual leather sheath. Cut off the tape holding your designs together and make it flat.

Step 3. Tracing and Cutting Your Leather Piece

Draw your pattern on the wrong side of the leather (the furry suede side), this is done because 1. it is not difficult; and 2. it structures the loop of the belt so that the right side faces forward. Usually, I overlook the belt loops section of the loop and only use it as a guide to trace along the piece using a ruler to ensure that it is straight and of adequate length.

Using a rotary cutter, make a cut-out of your leather but be careful so you do not cut into the inside edges where the blade part of the sheath joins the loop because by doing this, you will cut too much and make unattractive dents. Do not touch these areas; use only a sharp knife or an X-ACTO knife to complete the cuts.

Step 4. Begin to Create Leather

Using a lot of plastic wraps, wrap your preferred item (knife, compass, whatever) and tape it in to enfold it in properly. Lay out your dish towel, the item you want to sheath, a pan of hot tap water and the spring clips. Put the sheath part of your leather in hot water; as the leather absorbs the water, the color changes and you see bubbles. A few minutes is enough for this. Position your leather on the dish towel, fold the towel on the leather, and press down to dry it gently and mop off excess water. Position the knife on your leather and fold it over, pressing it along the handle as you progress. Firmly fix the leather into place using the spring clips; manipulate the leather, so it takes the shape of the blade handle. You can press the leather with your fingers to firmly attach it to the handle. Allow to dry. Though, check every five minutes for the first 30 minutes to ensure that the leather is molded to your satisfaction. Your leather can be used after several hours depending on the humidity and temperature, or you can choose to let it stay through the night. Remove the spring clips as soon as the leather is dry, and this would create a sheath 'husk.'

Step 5. Sheath Trimming and Preparing to Stitch the Seam

Trim the sheath to its actual size by cutting off the rough edges while following the outline of the blade and handle; to do this, use a rotary cutter. Since you are cutting through two layers of leather that are already hardened by water, you will need to apply more pressure. Do this carefully, so you do not cut yourself. Cut a shallow groove into the leather using the edge of the sheath seam as a guide, use a leather gouge to do this. It can also be done freehand or with a gouge that has an inbuilt guide. Using a rowel tool, mark your stitches in the groove. Six holes in every inch are enough. In the absence of a rowel tool, it can be done freehand, but this requires utmost care and precision. Position your sheath on a plastic cutting board and use your fid to make holes in the depression already made with the rowel tool. Tap your fid lightly with a small mallet. Once all the holes have been filled, pull up the top layer of the sheath and do the same for your bottom because your fid will have holes on the bottom layer as well. Ensure that they are properly aligned or your stitching would be stressful; in the absence of a fid, an ice pick, or any pointed object can be used. Fids are preferable because they create small slits and not holes.

Step 6. Sew up the Belt Loop

It is better to sew the belt loop now before stitching the sheath up. Fold your belt flap over to the front and correct it to your preferred size before trimming. It should fasten immediately below the top of the sheath. Once it is too deep, it may be difficult for the handle to properly seat in the sheath.

With the use of a four-prong punch, create an arrow of holes in the belt end of the loop and the top of the sheath. A fid, ice pick, or any sharp and pointy object can be used if you do not have a four-prong punch. If there is excess leather, trim it away from the end of the strap. Use your needle and artificial sinew to stitch the loop, going inward and outward until the stitches are visible. Tie your thread off and get your sinew cut close to the knot.

Step 7. Sewing the Seam

Use a sinew and a single needle to begin sewing from the lower part of the piece, close to the upper part of the sheath. Make a stitch with the side moving up through the leather and down into the next hole. A double-needle technique could be used, but since the seam is short, a single-needle technique will do. Once you have gotten to the end of your sheath, turn around and move up starting from the bottom, doing the reverse of what you just did. The effect is to make it tight enough to prevent it from loosing; the thread is protected with the groove in the leather and properly aligned with the leather or below the surface of the leather. Tie off the knots and thread the needle a few times inward and outward. Complete by threading the needle through a single layer of leather and draw taut. Cut off the lacing flush together with the seam to hide it. Use the wooden end of your fid to polish the seam of your sheath to even out the stitches then press it downwards into the sheath.

Step 8. Slide Your Knife In

Slide in your knife. It should be a little tight; it will loosen after a while. Position it on your belt. Take pleasure in the fact that you created something nice.

Sword Scabbard

Throughout history, scabbards were created (from middle, outward) from wood, leather, or wool. Then, two thin planks are also soaked and curved to take the shape of the blade. This is more time consuming, and with the tools I use, I would create a 'wedged' scabbard—two plain ones and one with a groove.

The materials and tools required are:

- Sword—Obviously, but not necessary. Every scabbard is customized, and the size is determined by the sword. If it isn't customized, it may not fit (sword might easily slip off and fall) or too snug (like the Excalibur-style scabbard) and can't be brought out once inserted.

- Wooden plank—The smallest size that can be used: blade length +5 – 10 cm, width of blade 4 cm. Adding excess length makes your scabbard more pointed. Make sure to select a plank that has the exact thickness of your blade or a bit less.

- Leather—has to be a lot bigger than the

wood plank because you will almost certainly cut your leather badly. The leather is also needed for the pattern and covering the pattern. I suggest that you buy leftover leather instead, as it is very affordable. You might want to buy leather laces to use it in strapping the bridge for instance.

- Wood piece for the bridge—size is determined by what you want

- Plenty of clamps

- Wood glue and brush

- Already waxed thread and leather needles.

- Leather glue (wood glue can be used if you want)

- Wood oil and finish—Select oil that is resistant to humidity and torsion, like outdoor furniture oil and wood flooring waxed oil. I had a leftover of flooring transparent wax oil, which worked fine. Clean the brush with petrol—better for oil brushes

- Special leather thread as the typical sewing thread isn't strong enough.

- Elastics

- Soap

- Ladder

- Pen

- 220-paper sander

- Jigsaw

- Mask and goggles

- Belt sander

- Tissue paper

- Leather hole punch

- Expandable towel

- Leather cutter

- Cutter mat

- Dremel

- Two buckets

Steps for Making a Scabbard
Step 1. Construct the Bridge

Although this step is not obligatory as a FIRST step, it MUST be carried out before Step 5.

Draw a pattern out to have on it some three-dimensional relief–to ensure that you will be able to place the lace pattern on the appropriate place, think of lacing and belt.

- To start with, draw out the holes you will like to have.

- You see its side on the schema.

- At least, you need a bottom hole that is big (to enable you to pass the strap that is used to tie the belt)—double holes on its upper part enables you to pass the laces and to tie it up on the scabbard.

- The double holes on the bottom side are not important. Create them in a way to get the lace on the scabbard surrounded; it can be solely flat.

- To cut off those holes, make use of whatever you want. A jigsaw could be used. A scroll saw is the best option if you have it.

- After you have done that, utilize the Dremel and belt sander to round the entire thing.

- Utilize mask and goggles

- You may have the desire to draw some art on it. I carved using a dremel. Also, you may pyro-draw; or do nothing, plain wood is okay as well.

- After that is done, consider treating the wood (oiling, varnishing, or anything else) and allow it to dry.

Step 2. Measure, Trace and Cut the Planks

- Place your sword on top of the plank, your traces ought to be 2 cm bigger than the width of the blade (1 cm on each of its side), and it should be 5 – 10 cm longer in length than the original length (this is dependent on how pointy you need the scabbard to be).

- Trace three nice rectangles out of those measured above (rectangles ought to be similar).

- Right at each rectangle's center, place the blade nicely (utilize the ladder to put the point accurately, a length of 1 cm on each of the guard's side), and draw out the profile of the blade. After this, fill the profile (to enable you to see where the blade is meant to be with less difficulty).

- Repeat these six times—three rectangles, two faces on every rectangle.

- Cut up the three rectangles.

- You will be required to cut the internal portion on one of the rectangles (the part where the blade goes). Because the sides are thin, they can disintegrate easily, so you have to be extremely careful (when using a jigsaw especially).

- It is okay if it breaks once or twice. You don't need to do it again (I would though, as a perfectionist), the issue might be fixed by gluing it all together.

- I just utilized a wood driller (25 mm) to cut the tip out to make a decent curve.

Step 3. Tryouts, Inner Scrape, and Glue Planks

Now is the time to be sure it fits your sword perfectly.

- You don't want it to be too loose or too tight, do not forget this. Clamp your three planks together (hole plank at the center) just like a sandwich without using any glue. Perfectly align them (and utilize the blade markup on the planks). Utilize plenty of clamps; the wood has to be very tight together so you should simulate the gluing.

- Make attempts to put in the blade.

- You will need to add something on the plain planks to add some thickness if it's excessively loose (it just keeps sliding in and off by gravity) Fabrics would be okay. Wool fabrics are even better.

- You will need to scrape a little of the plain planks to include some space in case it's excessively tight (you may have to push a bit though). Make use of a Dremel for that purpose—to make its internal portion smooth and to prevent damage to the blade, utilize the 220 sandpaper after that.

- Continue to repeat the process until it becomes tight enough. It will be a bit more tightened by gluing it together, so it must be a bit, just a little bit, not enough to tighten it to your taste.

- To clean the planks, a wet tissue could be used for this and take away all of the wood junk gotten as a result of this procedure.

- To get the planks glued together, you need to have the glue on the whole side, so you should use a brush.

- Glue one plank that is plain to the hole plank, clamp it together, use a tissue to clean the exceeding glue that is inside (to prevent making thickness with glue) and wait for 30 minutes till it's dry enough.

- Glue the last of the planks to the assembly and clamp it together. To be certain that the exceeding glue is not creating thickness, slide your blade in and out for about 3 – 5 times. After this, CLEAN UP YOUR BLADE!!! (WD40, gun oil, and so on)

- Wait till the glue is dry (read recommendations from the manufacturer of the glue)

Step 4. Outer Scrape & Oil Wood Scabbard

- On the table, fix your belt Sander and start scraping it to have an oval that is nice. You don't want to end up with any hole on the blade space, so TAKE AS MUCH TIME AS YOU NEED.

- Use goggles and mask

- Oil its outer part (add a great amount of oil, as much as possible, it ought to get into the wood's thickness).

- Allow it to dry up (read up recommendations from the manufacturer of the oil).

Step 5. Draw a Pattern to Have Some 3D Relief on It–Think About Belt and Lacing

- To draw out the pattern, use a leather strap of length 5 mm, they are directly glued on to the wood.

- At the barest minimum, to make two gaps to bind the bridge you will need to have four straps.

- A bigger spacer should be added a little more in the center, to have the strap which will be in charge of maintaining the sword diagonally on your hip.

- Then just for the outward appearance, add some pattern.

Step 6. Cut and Punch Holes on to the Covering Leather

- Cut up the covering leather. Cutting it up is tricky, the issue is that you don't want to have excessive leather, neither do you want to have it insufficient

- On each of the sides, punch holes evenly on them

Step 7. Prepare Covering Pattern

- You should set up a covering pattern. This is to be applied on the leather (step 9) once it's all sewed up and it is drying; this is done to have a nice 3D relief as the outcome.

- Consider your leather's thickness; it ought not to touch the glued pattern exactly, add a gap of +-2 mm (depending on the thickness of your covering leather).

Step 8. Soak Leather & Prepare Sewing & Gluing

- Make provisions for a basin of cold-ish water—present temperature

- Prepare another basin of water that has

been heated (not boiling water; something like a temperature between 40 and 60 degrees).

- For the entire leather soaking and molding process, if you want to know more about it, refer to Google. This procedure was flawless to me; it gave adequate time to sew and fix pattern before it's too hard, the leather was still soft enough after that (no need for it to be solid leather), etc.

- Soak up your piece of leather in cold water for ten minutes.

- Make provisions for a long waxed thread (it should be approx. 4 – 5 times your scabbard's length) and two different needles during those ten minutes. In case you end up with a thread that is excessively short (or you need it to be shorter to ease its sewing).

- You can cover up your scabbard of glue, once you are done with this. Make use of a brush to apply glue on all the wood.

- Once the preparation is done, the leather should have been soaked for approximately ten minutes now. Change your buckets, soak the leather inside water that is mildly warm for a period of 1 – 2 minutes (not more than this, the warmer the water is, the lesser the soaking time ought to be) and go on to the NEXT STEP (quickly).

Step 9. Sew, Apply Your Covering Pattern, and Allow it to Dry Overnight

- Pick up your leather piece, dry it up a little on your towel (its internal part, to prevent the dripping of water like hell on the wood) place it on your towel, outside and facing downwards on a table, place the scabbard on it and begin to sew.

- You should try to be fast (however, you should not be too fast, don't make errors) and, also preferably, on your sewing as well (if the diagonal, top left down to the bottom right is on the upper side on the first cross, it ought to be so all the time. Use blue thread on the external site).

- You should tighten it well, however, be careful not to break the leather. You should not hesitate to apply the leather very well with your hands, and if you must, stretch it.

- To prevent having problems with the thread, endeavor not to touch the glue too much with the needle or thread.

- NEVER USE YOUR FINGERNAILS, it will only leave bad traces.

- To stop the wire when you get to the throat, turn three crosses downward.

- You may now apply the covering pattern (to upgrade 3D relief & shape the leather) and tightly wrap it using elastics or cord, probably some clamps here and there. Allow it to dry overnight.

- Take off the covering wrapping & pattern in the morning and enjoy.

Step 10. Finish Throat and Chape (Leather Finish)

- As regards the throat, to enable the leather finishes to stay directly upon the guard, cut off the exceeding leather to leave a length of approx. 2 mm on both sides (thickness of the wood's edges). After this, add some glue and then paste the leather on it. To enable you to apply some pressure to keep it glued, slide in the sword so that the guard would create some pressure.

- As for the chape (the scabbard's ending), opt for one made of leather (or of metal). Just sew two portions together, soak it up for ten minutes, after this soak it in water that is VERY HOT (almost boiling) for two minutes (to make the leather hard). After this, apply it on the tip and allow it to dry overnight.

Step 11. Strap the Bridge, Belt or Baldric

To start with, you will lace the bridge on top of the scabbard, right on the placements which you have rendered before covering. Make sure that the lace is excessively tight; the bridge holds the straps that carry the entire assembly (sword and scabbard).

Chapter 7. Grip Materials and Scrimshaw

Just as the knife blades are important, Knife handles are also important. The knife will fail to function appropriately in the absence of a good handle. Due to the knife handle's importance, I have provided this information about the common styles and kinds of knife handles that exist. Handles of knives have been created out of pretty much every kind of material imaginable, ranging from the weird materials to incredibly nice looking and practical materials. Today, there are many artificial and natural materials that are nice for the creation of knife handles.

Grip Materials

The major purpose of the list below is to offer you some idea of the more common kinds of knife handle materials that are used in present times. You may likewise enjoy a list that is more detailed.

Abalone. This is a natural material obtained from a mollusk's shell. It is harvested off the coast of California, Mexico, and other regions of the South Pacific. Although it possesses an appearance that is extremely satisfying, it is not so durable compared to some other materials used for making knife handle. Its most widely known to be used as gentleman's pocket knives, here it will not experience exposure to the tumble and roughness outdoor use of a heavy-duty knife. Also, an imitation abalone is produced from a kind of plastic which is majorly used for the handle of pocket knives.

ABS. This is a black amorphous thermoplastic terpolymer, which possesses the strength of high impact. Resistance and toughness which are incredible characteristics for knife handles are the most astounding mechanical properties ABS possesses. Mostly, it is used for everyday working knives.

African Blackwood. An African Blackwood, also referred to as Mozambique Ebony, this is a rich black with dark brown graining. It is one of the extremely best woods for making handles of knives; it is used to produce fine clarinets.

Almite. This is a coating used on handles made from aluminum, like anodizing. It has resistance to scratching and marring. Also, it can be tinted to any color of your choice for it to be visually appealing.

Aluminum. Aluminum is a non-ferrous metal just like titanium is. It is commonly used as knife handles; aluminum provides the knife with a solid feel with the additional weight absent. Aluminum provides a solid feel for the knife's handle with the extra weight absent, and it is also very durable. It can be made to offer a grasp that is secure and comfortable. The T6-6061, a heat-treatable grade is the most common aluminum form. Anodizing is the most common finishing procedure for aluminum, and it adds color and protection to the handle.

Amber. This is a fossilized pitch from pre-historic evergreens, greatly utilized in jewelry; presently used by a few producers of handmade knives.

Ambidextrous. It is used with both hands with no difficulty. It is a knife which has not been exclusively created for a left- or right-handed individual but can be used with an equal amount of ease by the two hands.

Amboina Wood (spelled as Amboyna as well). Sometimes it is called padouk; it is an uncommon, exotic hardwood with a fragrant smell varying in color from red color to golden brown to yellow. It is utilized in cabinet production, and it is an incredible wood choice for both turning and finishing. It is obtained from the *Pterocarpus indicus* tree of the Southeast Asia jungles.

Anodized Aluminum. This is subjecting aluminum to electrolytic activity, which coats the aluminum with a film that is defensive and decorative.

Axis Deer (India Stag). The smaller sized of the SE Asian deer and the two Indian that provide antler for the knife production industry; all these are shed horn collected by the natives inside the jungle.

Bi-Directional Texturing. This is Spyderco's patented texture design molded into FRN handles with forward and backward graduating steps emanating outward from the handle's center. It gives resistance to slipping off from the hand.

Black Mother of Pearl. In today's knife market, this is one of the most sought-after exclusive pearls. It is obtained from small shells discovered in French Polynesia around Tahiti. Just beneath the exterior bark of the shell, there lies the Blacklip shell's real magnificence. From the standpoint of durability and utilization, it is extremely like abalone.

Bone. The bone which is utilized to make handles of knives is obtained from animals that died naturally. For improved grip and additional beauty, handles of bone are usually given a surface texture. To create the surface texture on the bone, jigging is the most widely recognized approach, and it is carried out utilizing a special machine for jigging in which modified bits cut out bone pieces. The machine works in a rocking motion to deliver the specific pattern that is wanted. Every one of the patterns possesses its distinguished look. The bone can be dyed in an assortment of colors after it has been jigged. Owing to the facts that bone is durable, fairly less difficult to shape and can be attractive, it is a very great material for knife handles. It is one of the most well-known handle material utilized for pocket knives.

Carbon Fiber. Fibers of Graphite (human hair sizes) are woven together and fused inside the epoxy resin. It possesses lightweight; it is three-dimensional in appearance and is a superior (and costly) handle material. It has a profound futuristic look with a definite "ahhhhh" factor. It is likely that Carbon fiber is the strongest of all the lightweight manufactured handle materials. The carbon strand's ability to reflect light, making the pattern of the weave extremely visible is the main visual attraction this material possesses. Additionally, carbon fiber is a labor-intensive material that produces a knife that is rather expensive.

Chital (check Axis – India Stag). The smaller of both the Indian and SE Asian deer that furnish antler for the knife business; all these are shed horn collected in the jungle by locals.

Cocobolo Wood. Hardwood obtained from the Cocobolo tree, and it ranges in color from dark purple to deep red and bright orange. The wood's grain and fine texture are relatively not difficult to work, shines to a high sheen and is well known as an inlay or embellishment on handles of knives.

Cordia Wood. This wood type is extremely similar to Teak, and occasionally, it is utilized as a substitute for Teak when it comes to building ships.

Desert Ironwood. This wood is native to the Sonoran desert in southern Arizona and Northern Sonora Mexico. The wood is exceptionally dense and tight grained, it takes an extremely high polish, and tends to darken with use and age.

European Stag. Antler obtained from the Red Deer, a big elk-like creature found all around Europe. For at least as long as knives of metal have existed, and likely a long time before then, it has been utilized to make handles of knives. Never has this stag been a substitute for the axis and sambar deer of Southeast Asia and India's antlers. Much like the American elk, the European Red Deer's center is exceptionally coarse and open. Mostly, it has to be utilized as handle scales due to the expansive measure of pith in the center. The Red Deer's antler is a limited substitute for both the Axis and the Sambar's antlers.

Fiberglass Reinforced Nylon. A nylon polymer combined with glass fiber which is then injected into a mold to create lightweight handles of knives.

Forprene. Forprene is a material that is highly resistant, an elastopolymer with extremely high thermal properties from a temperature of -40 Degrees C to +150 degrees C; it has exceptional high-grip power. Also, the material is exceptionally salt and acid corrosion- resistant and can be utilized in every single wet situation.

G-10. G-10 is a laminate that is fiberglass based. Layers of fiberglass cloth are soaked inside resin, compressed and then baked after this. The resulting material is hard, lightweight, and solid. The surface texture is included in checkering form or other patterns. Because it is durable and lightweight as well, it is a perfect material for tactical folders and fixed blades of knives. It comes in different colors.

Glass Filled Nylon. A great number of today's thermoplastic materials are improved on by adding chopped glass strands. Regularly, as much as 40% of a product's percentage might be glass. It adds incredible strength.

German Silver. It is a combination of copper, zinc, and nickel. It is otherwise referred to as Nickel Silver.

Jigged Bone. Obtained from dead animals, mostly a cow's chin bone. Generally, the bone is dyed, and by cutting grooves into the bone, surface texture is acquired. It was first utilized in the imitation of genuine stag scales.

Kraton. This is a thermoplastic polymer which is rubbery and is utilized for the handle of knives or as a flexible inlay on knife handles for an improved grip.

Laminated Handles are handles created from different materials that are layered together and held together by glue.

Leather. On some hunting and military knives, leather handles are present. By stacking leather washers or as a sleeve encompassing another handle material, leather handles are typically created. Although knife handles made of leather are attractive to look at, they are not as durable as some other materials. As spacers to add accents to the handle of a knife, leather functions admirably.

Mammoth Bone (Molar and Ivory as well). Rarely utilized in custom knives. It is found during mining tasks in the far north, in locations with plenty of glacial activity. The distinguishable look is created from erosion.

Micarta. When it comes to construction, Micarta is similar to G-10. Either canvas, linen cloths, or paper layers are soaked inside a phenolic resin. Pressure and heat are applied to the layers which ignite a chemical response (polymerization) to occur. The final product is a material that is lightweight and strong. Because of its extraordinary toughness and stability, Micarta is a well-known handle material on user knives. Micarta has come to be known as practically any fibrous material placed in resin. It's available in an assortment of colors and laminate materials. Micarta is very smooth to touch; it does not possess surface texture. This material demands hand labor, which implies a higher valued knife. If not treated appropriately, Micarta is a material that is relatively soft and can be scratched.

Mother of Pearl. The pearl oyster's shell from the South Pacific; a knife handle material that is costly and well known.

Natural Materials. Natural materials, for instance, jigged bone, mother of pearl, abalone, leather, stabilized woods and stone that are utilized in creating and embellishing handles.

Palmira Wood. Kitul Black Palmira possess a unique structure with dark, nearly black, tough streaks in a background matrix that is paler. Although it is hard to work, it produces results that are very dramatic.

Pearl. The pearl oyster's shell from the South Pacific; a costly and well-known knife handle material.

Peel Ply Carbon Fiber. It is a carbon fiber filled, epoxy resin lay-up that has textured material set superficially to protect the material while manufacturing is going on. The material is removed after production, and it leaves a texture that is grippy in the epoxy creating a non-slip handle material.

Polycarbonate. A strong synthetic resin utilized in molded items, for instance, handles of knives, unbreakable windows, and optical lenses.

Sambar. An extremely large, elk-sized deer found in India and S.E. Asia; the antler is utilized for knife handles and is referred to as or India stag.

Sermollan. A plastic that is rubberized and utilized on kitchen knife handles. It offers a secure grasp and is resistant to bacteria.

Stag. Stag is also another material that is very popular. Because of its high density when compared to others, Sambar Stag antler material is the most sought after for creating knife handles of all the species of deer. Most Sambar Stag is from India and due to the ban by the government on its export, it is becoming increasingly expensive and uncommon. Stag is obtained from naturally shed deer antlers. Stag takes on that slightly burnt look when exposed to open fire. Sambar Stag makes superb knife handle material, and it's extremely durable.

Stainless Steel. Is referred to as steel that contains at least 12-1/2-13% chromium, making it corrosion-resistant (it is not stain-proof). The chromium oxide CrO prevents the formation of rust by creating a barrier to oxygen and moisture. Although various grades of stainless steel exist, nearly all stainless steel blades contain a great amount of high carbon, so none are entirely "stainless." All grades are liable to corrosion from humidity, body acid, salt, etc. The term has come to imply that the steel has a fewer amount of carbon and more amount of chromium, and will, therefore, stainless than the majority of other steels. To make it less difficult to grip, it is regularly utilized in combination with a different material such as rubber or plastic. The weight is the greatest drawback to the handle of stainless steel knives.

Titanium. Titanium is a metal that is known to be harder but lighter when compared to steel. While stainless steel handled knives are more often than not on the heavy side, titanium gives the toughness and durability of a metal handle that does not have so much weight. Of any steel, Titanium provides the most resistance to corrosion. It possesses a decent "feel" and makes a fantastic material for knife handles.

Valox. A handle material obtained from reinforced resin.

Volcano Grip. This term is Spyderco's trademark for the waffle texture found in their fiberglass reinforced nylon handled lightweight knives. While cutting, the continuous pattern of small-sized squares offers a better hand grasp.

White Mother of Pearl. A highly valued knife handle material! It originates from the silver lip shell. Few of the best White Mother of Pearl originates from tropical Australia's South Seas. It is extremely uncommon in sizes large enough to utilize for knife scales. It is known that 10 tons of pearl shells are required to obtain material large enough to cut sizes 1/10" x 4-1/2" long. From the angle of durability and advantages, it is also highly similar to abalone.

Wood. Wood handles of knives vary from the more typical wood species to the most exotic species, and the value ranges accordingly. For hunting knives or for uses that involve a great deal of moisture or water, soft or fine woods such as black walnut are bad choices. Hardwoods such as Rosewood, oak, and maple are good choices for making hunting knives. Stabilized wood such as mesquite, desert ironwood, and spalted maple are available where the wood is infused with plastic making it waterproof and furnishing it with a durable finish which does not need any maintenance apart from buffing it occasionally. For tough duty knives and knives that would come in contact with a great deal of moisture, these are highly recommended. Exotic and Fancy knives with wood handles are particularly popular with collectors. A wood handle that is of good quality will be durable and can be attractive, too.

Wood Epoxy Laminate. It is an impregnated wood laminate, which is very hard and machines in a way that is similar to Corian, aluminum, and Micarta.

Zytel. This thermoplastic material was developed by Du Pont. ZYTEL is the most economical to create of all synthetic materials. It cannot be broken: it resists effect and abrasions. ZYTEL possess a slight texture in regards to its surface, but knife companies utilizing this material will include extra, aggressive surface texture to in addition to the slight texture it possesses. For utilizing Zytel, SOG Specialty Knives is not rare.

Numerous other materials utilized for making handles of knives exist as well as different kinds of plastics and exotic materials for instance: warthog tusks, mammoth tooth, and ivory, oosic (walrus penis bone), stone, sheep and buffalo horn, etc. Practically all hard material can be (and has been) utilized as a knife handle.

How to Scrimshaw

Scrimshaw is a classic American folk art form that was once perfected by sailors from New England. They would make designs onto whale ivory using needles or knives and color it using ink or soot. Although the whaling trade is not viable or legitimate anymore, the scrimshaw craft lives on.

Part 1. Finding Materials to use

1. Check ancient salvage yards or thrift stores for small sized ivory pieces. If you are utilizing whale ivory, ensure that it was harvested before the year 1972 when the Marine Protection Act banned whaling in America. You can likewise utilize ancient keys of an ivory piano,

bone or white acrylic.

2. Purchase a pen-like X-ACTO knife that has a replaceable head. Introduce a pin into the front and secure this pinhead inside the knife.

3. Obtain some beeswax, black, brown or blue ink and some acetone nail polish cleaner, all of good quality.

PART 2. Sealing the surface

1. Rub beeswax to the uppermost of the ivory or bone to seal it, if you can, put a disk to a Dremel tool and rub the beeswax on the disc, too. Again, rub the beeswax equally to the topmost area of the ivory, covering it. Why are we sealing the ivory if you may ask? This is done because ivory is penetrable. This sealing will help keep the ink that is tattooed into the ivory away from dropping over into where it isn't needed, leaving an inky cloud. If you properly sealed, the ivory will take up only the ink that went into the grooves that were etched into it.

2. Use a dry cloth to spread the beeswax into the surface by hand for like 5

minutes especially if you do not have the needed tool to work with, try to apply the beeswax all over the ivory evenly until it gets to all part of it

3. Using a clean cloth next, polish the ivory until all the beeswax has been taken off. The ivory should be very sparkling at this point but no longer glistening, keep the beeswax cloth aside since you will still need it later

Part 3. Transferring a design

1. Measure your piece of ivory; you will need to use your scrimshaw as your small detailed drawing

2. Look for an image online and resize to the size of your object, ensure to leave one-half inch of space on all sides. A really good draft out image will perfect outlines, and maybe some shadowing will be helpful with scrimshaw.

3. Either you print the image using a computer, or you copy it from a book to a white paper.

4. You can now place the ivory on top of it and draw an online of the ivory over the white paper, or you cut it to straighten it

with the edges swiftly.

5. Now place the illustration face down on the paper, dampened a clean cloth with acetone nail remover and apply it gently over the top of the paper using the cloth and then again with a bone folder. Ensure that the paper is moist.

6. You can now lift the edge of the paper and remove it back very fast. Don't try moving it around on the exterior to avoid smudging your outline. You can throw away the paper now. Check it out if the edge didn't come off on the bone visibly, then try to sand it, maybe wax it and redo it.

Part 4. Etching the surface

1. Use your pen to draw the outlines of the illustration, start by applying pressure withholding the pin vertically as you can. Etch the lines into the surface of the bone.

2. Complete the outline of the illustration, next, rub some ink on the surface with a cotton swab, rub a good quantity and then clean it off the excess with a lint-free rag

3. Start shading the drawing by etching crosshatched lines. Also, you can stipple by etching dots because the closer the dots to each other, the darker the shading will become.

4. Spread more ink on the cotton swabs and clean it off, note if you need a darker line still, you can draw it deeper and pour more ink on it.

5. Replace your pins if need be when once you noticed it begins to dull

Part 5. Finishing the scrimshaw

1. Recheck your work to identify a mistake immediately if any, then draw larger lines or you sand the area, you can also re-wax or sketch if the mistake needs any of these.

2. Clean all the extra ink away from the surface of the bone

3. Get your beeswax covered cloth, then polish the surface of your scrimshaw when completed by spreading an equal layer over the artwork. This should be sparkling and save the ink.

Chapter 8. Heat Treating the Blade

A good knife will never dull with little use or break with little pressure, but the same can't be said about a bad knife. A good knife will remain sharp for a long while and remain straight and true even with a whole lot of stress on it. This to a large extent is the difference between knife sold in Walmart and a custom handmade knife. With various heating and cooling, steel changes physical properties, your target is to manage these properties to suit the use of a knife. Though you might have heard about this before now, it's also good to get a good knowledge about it proper workings.

The Two Physical Properties

When it comes to knives, there are two needed properties. The first one is a knife hardness and brittleness. A hard knife is likely not to dull, it is also likely to be used a lot on various materials, and the cutting edge will remain solid. Well, a hard can snap easily though. Primarily, the atoms are formed in a rigid pattern; the atoms would rather break bonds than completely move slightly in addition to one another.

The second property a knife should possess is toughness and softness. A good tough knife can be used even when hunting, and it will remain the same. The only problem will be the cutting edge not staying sharp for too long because of its softness and of course it will become a piece of metal that stays in a wedge shape. But on the molecular level, instead of the atoms breaking bonds totally, they would move slightly about each other, this is to say that the cutting edge microscopically folds over itself. This can be seen as a sliding scale, one end is hard, and at the other end, it feels tough. Your target should be to get the right stability of the two as hard as you can and also make it remain sharp as long as you can. But you should still be careful not to make it too hard that it will get easily broken under some reasonable pressure. It should be able to stay for a long time. Going to the actual process the step by step heat treating will be discussed, and a molecular explanation will be given as well. It will be great to know the science behind this so that it will help in resolving issues if you ever run into any of them

Equipment and Materials:

- Tongs or visegrip pliers
- Forge

- Oven
- Fireproof quench container and cover
- Quenching medium like motor oil
- Fire extinguisher that puts out oil and grease fires
- Heat resistant gloves and face shield
- Fireproof material for regulator block such as an aluminum tube
- One forged or stock removed high or mild carbon steel knife blank

Heat Treatment and Etching

Process of Damascus

Materials needed: 1095 and 15N20

Process

1. Heat the blade to between 760 and 790 degrees C meaning critical temperature

2. Quench in light or quenching oil

3. Lastly, drag temper to a possible hardness

Heating

The blade should be gently and evenly heated to a bright red or dull orange color about 760 to 815 degree C should be used, this likes heating up to about 15-20 minutes. Ensure you do not overheat it as an overheated blade can become cracked or warped when quenched.

Hold at the critical temperature for up to 3 minutes. You can as well try this other good method, of heating the blade till it loses all its magnetic properties. This usually happens at about 770 degrees C. If you, therefore, want to ascertain that a blade is ready for use, you can quench it with a magnet. When there is little attraction or none at all between the magnet and the blade, be sure that the ideal temperature has been reached and is ready to quench.

Quenching

Quenching can conveniently be done in either the standard quenching oil or in light oil. Oil quench can be used on a large blade where toughness is needed more while brine quench is needed more on a smaller folder and skinner blades where holding an edge is very important. Brine quenching can be used for smaller blades too. This will help make the blade be very hard as it will assist in cooling the steel much quicker than even the oil. But you have to be careful when quenching in brine. There is a higher likelihood of the blade cracking if it's cooled immediately. The best bet is to preheat the brine to about 100 degrees before quenching. Dissolving salt in boiling water until it no longer melts makes a brine solution. The thin blade needs to be quenched either point first or its spine first, this is to help reduce the opportunity of cracking or warping. To quench the thick blades then the cutting edge should be quenched. First, this will assist in ensuring hardness on the cutting edge and don't forget to keep the blade moving while quenching.

Tempering

Tempering the blade is all about heating the steel in a heat treatment oven. This should be done immediately after quenching to reduce the opportunity of cracking which could stem from remaining stresses. Leave to remain at that temperature to ascertain an even full heat. The excellent way to get the same temper is to heat a bigger size of slab or block of steel to a specific temperature, and then set the blade on that and leaving it there to absorb the heat. Leave it at the tempering temperature as long as you can maintain it or maybe up to an hour. If you had used a large tempering block, then let it be so that the blade will cool together with it. The relative hardness of the blade is dependent on the specific temperature. The recommended drawing timeline should be two hours. Don't forget to leave enough space for each blade to aid proper air circulation.

Deg C	As Quenched	
205	260	315
Hardness (RC)	63-65	58-59
	56-57	53-54

Heat treatment for stainless steel

Stainless steel is primarily divided into four groups to know their reaction to heat treatments. The precipitation hardening grades, assigned a PH suffix e. g 15-5 PH consists of a group which like the other three achieve both high strength and corrosion resistance. The following is the short notes of the heat treatment potential of the two groups

Ferritic stainless steels: this stainless isn't hardened by quenching. The heat treatment used for ferritic is annealing since they have minimum hardness and maximum flexibility, impact, toughness, and corrosion. Annealing helps to reduce stresses which were generated during the welding or cold working. It helps to provide a better uniform microstructure.

Austenitic stainless steels

The normal austenitic stainless steels will be hard when working with cold but can't be the same when working with heat treatment. Annealing is used to optimize softness, corrosion resistance, and ductility. Nevertheless, post-annealing can be specified after thermal processing or welding. This group is usually bought in the cold worked or annealed condition.

Martensitic stainless steels

The greatest hardness one can achieve when using martensitic stainless steels depends on the carbon content. So, therefore, heat treating this group is the same as for carbon or low alloy steels. The process parameters are different though since the higher alloy content of the group makes them react more slowly. They sometimes show absolute hardenability so that the highest hardness is achieved in the center of the sections which will be up to 30mm that is about 12 inches thick by cool airing. Conclusively, there are two main types of heat treatment.

Annealing helps to lessen hardness and enhance flexibility. Going full annealing is quite costly and take up time too, so it should only be specified when needed for severe forming. Austenitizing quenching and tempering are used to enhance strength and hardness. These methods also affect corrosion resistance; sometimes it needs balancing the heat treatment parameter to optimize corrosion resistance and product strength requirements.

Precipitation hardening

Stainless steels are the most basic forged grades of **precipitation hardening** (PH), stainless steels are a 17-7 PH and 15-5 PH. These grades come together to form high corrosion resistance of austenitic grades with the energy attainable in martensitic grades. There are available procedures for homogenization, austenite conditioning, transformation, cooling and precipitation hardening.

Duplex stainless steels: this is the main forging grades which contain the mixture of both ferrite and austenite in their microstructures. They are always not heat treated other than annealing.

Leave A Review?

Throughout the process of writing this book, I have tried to put down as much value and knowledge for the reader as possible. Some things I knew and practice, some others I spent time to research. I hope you found this book to be of benefit to you!

If you liked the book, would you consider leaving a review for it? It would really help my book, and I would be grateful to you for letting other people know that you like it.

Yours Sincerely,

Wes Sander

Conclusion: Time to Step Up

While there are a lot of books on bladesmithing, there are only a few books on in-depth knowledge of bladesmithing. With the level of renewed interest in knife and sword making, this book revealed all a blade enthusiast needs to take their learning to the next level. Having known all about the advanced methods of bladesmithing, it's time to step up from being a beginner. You have all you want now. This book contains knowledge you cannot get anywhere.

www.ingramcontent.com/pod-product-compliance
Lightning Source LLC
Chambersburg PA
CBHW070704190326
41458CB00004B/838